The MIT Press
Cambridge,
Massachusetts,
and London,
England

36 Lectures in Biology

S. E. Luria

Copyright © 1975 by
The Massachusetts Institute of Technology

This book was typed by Kathy Lane. It was printed
and bound by The Colonial Press Inc. in the
United States of America.

Library of Congress Cataloging in Publication Data

Luria, Salvador Edward.
 36 Lectures in Biology.

 Bibliography: p.
 1. Biology. I. Title. [DNLM: 1. Biology.
2. Genetics. QH308.2 L967t]
QH308.2.L87   574   74-19136
ISBN 0-262-12068-2
ISBN 0-262-62029-4 (pbk.)

CONTENTS

The lectures in this book are the substance of a
General Biology course taught at M.I.T. in the
springs 1973 and 1974. Most of them are based on
a transcript made in 1973. Five lectures that
were presented by other teachers in 1973 are included as I presented them in 1974.

The justifications for publishing this set of
lectures in book form are, first, that students
seemed to enjoy them and profit by them; and
second, that this course may offer a useful para-
digm for biology courses of a kind suitable to
good students who like to look under the surface
of the subject matter.

The course is centered around one main theme,
that of living organisms as possessors of a pro-
gram--a set of genetic information--that under-
lies all vital functions and that evolves by
mutation, genetic recombination, and natural
selection. This theme is followed through the
discussion of the chemistry of living cells,
their metabolism, and their synthetic activities,
which are directed essentially to the reproduction of the program. Genetics is the natural
center of the presentation. Developmental
biology and physiology are approached as the
expression of the program.

The question is bound to be raised that many
areas of biology--zoology, botany, evolution
theory, ecology--are not touched upon in these
lectures. My opinion has always been that these
disciplines, if they are to be learned well by
the science-minded student, are best approached
after a general biology course in which the
student learns the essentials of the life phenomena: the chemistry of the cell, the organization and function of the genetic systems, the
genetic significance of life cycles, and the
functioning of cells in differentiated organisms.
No claim is made that the lectures collected in
this book cover exhaustively even those areas of
biology that are included. The lectures represent a selection of topics organized around the
central theme of the biological program of
organisms. Thus some of the physiological topics,
for example, are described without special concern for the underlying cellular or organ structures.

The M.I.T. students are not required to take a
chemistry course as a prerequisite to this biology

course, but most of them are well read in all
areas of basic science. It has proved useful to
provide the student with some summaries of basic
facts of organic and physical chemistry needed to
understand elementary biochemistry. These summaries are included in the book as background
materials (pages 316B-402B). I have also included
a set of suggested topics for discussion at the
weekly recitations that accompany the lectures,
as well as a few sample examination questions to
illustrate the level of performance expected of
good students in this course.

The students have used as texts two books:
Watson's Molecular Biology of the Gene plus any
one of several biology texts of their choice. I
have listed at the end of this Preface a few
books that would provide adequate supplementary
readings for the eager student and have suggested
desirable readings at appropriate points in the
series of lectures.

In a long list of acknowledgments, the most
important one goes to Carole Bertozzi, who in the
fall of 1972, with financial support from the
Division for Study and Research in Education of
M.I.T., helped me collect materials for the
course and in the spring of 1973, with the help
of Eva Aufreiter, recorded and took notes of my
lectures. Carole then prepared a typescript that
served as a basis for my further work. The
excellence of her work made my task easier and
swifter. I have endeavored to preserve to some
extent the flavor and casualness of the lecture
room at the price of improving style and syntax.
I have also tried to give the drawings in this
book the quality of blackboard sketches.

I am indebted in various ways to those of my
colleagues who have participated in teaching
General Biology at M.I.T. in the last 10 years:
Paul Gross, E. C. Holt III, Vernon Ingram, and
Cyrus Levinthal. Even though there has not been
a great deal of explicitly joint planning, I am
grateful to John G. Nicholls and Stephen W. Kuffler
for a critical reading of two lectures. I am
also grateful to the numerous graduate students
who assisted in the teaching of the course. Nancy
Ahlquist prepared all the diagrams from blackboard
or pencil sketches, besides being involved in many
critically important ways in the preparation of
this book as she has for 15 years been involved in

every aspect of my writings.

Finally, my gratitude goes to my students, to whom this book is dedicated. I can think of no greater pleasure for a teacher than to see a crowd of bright faces respond with understanding to his lectures. This more than anything else encouraged me to put these lectures into print.

At the end of each lecture or group of lectures reference will be made to specific chapters in one or more of the following books as well as to specific parts of the Background Section of this book, pages

J. D. Ebert and I. M. Sussex. Interacting Systems in Development, 2nd Ed. Holt Rinehart and Winston, New York, 1970. An elementary but informative presentation of development with emphasis on cellular and molecular aspects.

B. D. Katz. Nerve, Muscle, and Synapse. McGraw-Hill, New York, 1966. A stimulating discussion of the central issues of neurobiology. Not easy but rewarding.

A. L. Lehninger. Biochemistry. Worth, New York, 1970. Undoubtedly the best biochemistry textbook available today.

I. M. Klotz. Energy Changes in Biochemical Reactions, 2nd Ed. Academic Press, New York, 1967. This delightful little book is not easy reading but is an excellent introduction to the fundamental concepts of chemical energetics.

G. S. Stent. Molecular Genetics. Freeman, San Francisco, 1971. This book deals with molecular biology from a narrative point of view, explaining the experimental basis of the various landmarks.

M. W. Strickberger. Genetics. Macmillan, New York, 1968. A very complete text of genetics that covers with equal competence formal genetics and population genetics.

A. J. Vander, J. H. Sherman, and D. S. Luciano. Human Physiology--The Mechanisms of Body Function. McGraw-Hill, New York, 1970. An elementary physiology text, deficient in biochemical analysis but appealing to a modern biologist because of the emphasis on regulatory mechanisms.

J. D. Watson. Molecular Biology of the Gene, 2nd Ed. Benjamin, New York, 1970. A book that deserves its fantastic success because of its superb intellectual clarity and its insights into the central issues of present day biology.

Part I

CELL BIOLOGY AND CELL CHEMISTRY

Lecture 1

This is a course in biology. Biology studies those entities that are called organisms: men, worms, yeast cells, bacterial cells are organisms. Some organisms are <u>unicellular</u>, some are <u>multicellular</u>. An organism is living; it is an individual that can grow and participate in reproduction generating similar organisms. To grow and reproduce, an organism uses materials from the nonliving environment to make "living substance;" that is, it assimilates. Some organisms, for example bacteria, can reproduce themselves by division; others, such as human beings, reproduce only by mating.

How does biology differ from other sciences, such as physics or chemistry? Physics studies the properties of matter and energy. The subject matter of physics is unchanging and the same physical laws are applicable now as have been applicable since the beginning of the universe some 15 billion years ago. (Cosmologists are not sure what happened before that, but 15 billion years is good enough for us.) Chemistry studies the approximate physics of substances composed of identical molecules. The study is approximate because quantum mechanics is not yet good enough to make it exact. As in physics, the subject matter of chemistry is assumed to be fundamentally unchanging even though organic chemists make new products every day.

The subject matter of present day biology, however, is not the same as 15 or even four billion years ago; in fact, life on earth did not exist at all then. Organisms present today are merely a sample of all the organisms that have existed since life began on earth. They are those relatively few groups of organisms that have managed to struggle and stay alive and produce descendants. Present day organisms thus include only a small subset of all those kinds that <u>might</u> have been (just as human history includes only a sample of all the events that might have occurred).

What distinguishes organisms? First, organisms <u>evolve</u>, that is, the types of organisms that exist change with time; second, organisms are unique in nature in that each organism has a <u>program</u>, the genetic material, which determines how the organism develops and functions and responds to various environments. An egg, for example, has a program that directs its development into a chicken. Development in a certain direction is controlled by instructions contained in the organism; these instructions compose the program.

Plan versus program

A program is different from a plan. A plan has a purpose or goal: the motions involved can be changed at any time to achieve that goal. For example, I can change my approach to teaching a course if the students become confused, in order to reach my goal of communicating certain information. A program, on the other hand, embodies no possibility for purposeful change. A computer programmed to play chess, for example, always responds in the same way to the same chess situation. A plan is creative, a program is passive.

Like the computer, the programmed organism has a tape containing the instructions (genes, DNA) and the machinery to implement the instructions on the tape. In theory at least, it is possible to extend the analogy one step further and build a computer than can reproduce itself, a so-called "von Neumann machine" (from the name of John von Neumann, a famous mathematician), that is, a computer with a tape that directs the machine to build another computer, which then in turn copies the tape. The machine could evolve, that is, change with time via mistakes or mutations made in copying the tape. Usually such mistakes would result in a machine that works less well than the original. Occasionally, however, the result would be a machine that works a little better. Or a machine that has suffered a harmful mutation may be helped by a change in environmental conditions (for example, the air conditioning in the building may go off); under a new set of conditions, a machine with a mistake in its tape may work a little better than the original.

Evolution

There is a strong analogy between this situation and the evolutionary changes that occur in organisms with time. Within a population of a given species of organism, variants arise by genetic mutations, which are changes in the structure or arrangement of the <u>genes</u>. These are the units of hereditary material that determine the hereditary properties of all organisms. Variety also arises upon reproduction by mating since at each generation the genetic material of the parents is reorganized, producing many new combinations of genes. Evolution occurs because

natural selection favors, among all the combinations available, those individuals whose characteristics will increase their reproduction in a
particular environment.

There is no good or bad in evolution; only
successful or unsuccessful. An organism that
functions well under certain sets of conditions
and therefore has more descendants is successful.
Under different conditions the same organism
might be unsuccessful. Thus evolution gauges in
a blind, purposeless way which organisms will
survive.

The theory of evolution by natural selection
is the fundamental generalization of biology. A
consequence of this theory is that all organisms
are related to one another because they share
part of a common descent, that is, some common
ancestry. Evolution can be represented as a
tree with certain lines dying out:

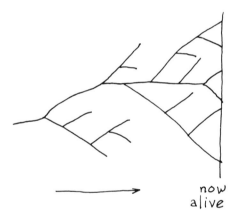

now
alive

Most of the species of organisms in the overall
tree of descent have become extinct.

**Unity of
biochemistry**

Another great generalization of biology is the
unity of biochemical processes: all organisms
now existing share certain basic biochemical
reactions. This is because all of them have two
things in common: the chemistry of the tape is
always the same, being made of nucleic acids, and
the machinery--proteins and devices for producing
proteins--is by-and-large always the same. That
is, the unity of biochemistry reflects the similarity of macromolecular types in all organisms.
Different types may have existed in early stages
of life's development on earth, but one single
successful line of descent, based on nucleic

acids and proteins as we know them, obviously took over quite early. Because of this, we can study certain fundamental processes of life in whatever organism is convenient. For example, the genetic code in bacteria is applicable to man. (The genetic code, as discussed later, is a set of correspondences that says that a particular sequence of symbols on a DNA tape is translated into a particular sequence of symbols in protein.)

Cell theory

The third central generalization of biology is the cell theory: all organisms consist of cells. (Viruses are not cells, yet they are not an exception: they are organisms that utilize the cells of other organisms as a cuckoo uses the nest of other birds.) Although still referred to as the "cell theory," this principle is really a factual generalization. The cell is a bag, a closed domain within which the sequences of chemical reactions needed for life are carried out.

Why make a cell? It is like asking what you gain by having a home. The cell is an invention for keeping the concentration of essential materials high enough so that the chemical reactions needed for life can take place at near optimal rates even if the external concentrations are too low or too high. This is because cells are surrounded by membranes that can selectively retain and even concentrate chemical compounds. Thus essential nutrients can be pumped into a cell. For example, the high concentration of potassium ion K^+ in many cells is maintained by a membrane pump that carries K^+ ions into the cell and keeps them there. Other substances are actively expelled from the cell by pumps in the cell membrane. The membrane also serves to exclude poisonous materials from the cell. Such careful regulation of in- and out-flow allows the genes and the rest of the cell machinery to function at high efficiency and is most probably the reason that cellular organisms in the course of evolution have displaced any noncellular form that might have evolved.

The most general approach to organisms is to study their compositions and organization. We wish to know first what chemical compounds and structures they contain and then learn how these are made (biosynthesis). We are particularly con-

cerned in tracing the program substances and
understanding how they work. We need to choose
some prototypes: these may be multicellular or
unicellular organisms. In these lectures we
choose as main prototypes a bacterium, Escherichia
coli (or E. coli), and a mammal, Homo sapiens. We
shall make reference to several other organisms,
including, for example, Neurospora (bread mold),
Drosophila (fruit fly), echinoderms (like the sea
urchin), and flowering plants. All are composed
of cells. E. coli is a unicellular organism;
each cell is about 1 μm^3 in volume and weighs
about 10^{-12}g. Man, a multicellular organism, is
composed of many different kinds of cells. A
liver cell, for example, is about 1,000 times the
volume of E. coli (1,000 μm^3) and weighs about
10^{-9}g.

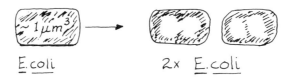

E.coli   2x E.coli

Procaryotes and eucaryotes

E. coli is a procaryote, man is a eucaryote.
This is an important distinction: procaryotic
(that is, prenuclear) organisms have cells much
simpler in organization and are therefore easier
to study. Eucaryotic (that is, well-nucleated)
organisms have complex cells with nuclei and
other internal compartments. Bacteria are procaryotes; all plants and animals are eucaryotes.
Procaryotes have taught us a lot about genes and
their functions, but one must be careful not to
extrapolate too freely from pro to eu!

As seen in a good electron micrograph, bacteria have just a wall, a membrane, and some less
opaque areas inside, where most of the DNA is
located. These inner areas are less electron
dense because they are more hydrated; that is,
they contain more water, which is removed upon
drying prior to electron microscopic examination.
Procaryotic cells are rather plain. Eucaryotic
cells, however, can be very beautiful. A liver
cell, properly examined, reveals, within a cytoplasm that is the viscous matrix of the cell, a
nucleus with a membrane interrupted by many pores;
a large number of mitochondria, where oxygen is
used to get energy out of a variety of foods; a

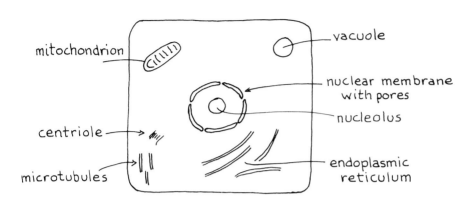

network of inner membranes where proteins are
made and other chemical reactions performed; and
additional structures related to the intake and
output of substances that the liver cell proc-
esses--the liver is the most diversified chemical
factory of our body. Later we shall mention
other cell types, each with its own function reflected in its structure. The most fantastically beautiful are the ciliate and flagellate pro-
tozoa. But for lack of time we shall not deal
with them in this book.

Here we may note that in the nucleus of our
liver cell we don't see the chromosomes. These
become visible only at mitosis, that is, at the
time of cell division; and liver cells do not
divide any more in the grown-up individual.

Our study of organisms will be geared around
the central concept of the program. We shall
seek answers to the following sorts of questions:
What is the program? How does one analyze it?
How does one grow cells and analyze the molecules
they make? Which molecules are used in the cell?
How do the molecules made by cells function? How
is their chemical structure related to their
function?

How are the program substances organized in
the cell? How are they transmitted from generation to generation? How are they reshuffled?

How does the program unfold? How does an
organism like E. coli manage to express different
parts of the program depending on which medium it
finds itself in? How does man manage to express
different sections of the program in different
cells of the body, such as liver cells or brain
cells? In other words, how do genes function in
the process of differentiation?

Cell growth

First, in order to study cells, we must have some cells to study. Both procaryotic cells, such as bacteria, and cells of eucaryotes, such as those in man, can be grown in cultures of an appropriate medium. It is considerably more difficult, however, to grow eucaryotic cells in culture than it is to grow bacteria. Human or mouse cells require complex mixtures of all sorts of nutrients, such as vitamins, that they are unable to manufacture. Much of the stuff that animal cells need is not known yet and one must provide it by adding blood serum to the culture medium. Also, in culturing cells from an organ such as the skin one has to digest the stuff that cements the cells together. Cancer cells are often easier to grow than normal cells.

E. coli, on the other hand, requires only some salts plus sugar or some other carbon compound and, of course, water. One liter of a good culture medium can yield more than one gram of E. coli cells, which can easily be collected in a centrifuge tube, washed, and submitted to all sort of indignities.

If we have in the laboratory a culture of cancer cells, we can compare the growth of E. coli cells and cancer cells in culture. We inoculate the appropriate liquid medium with a sample of cells, incubate, say, at 37°C, and remove samples at regular intervals to assay for the number N of viable cells per milliliter of culture medium. We obtain a growth curve by plotting N or log N versus time.

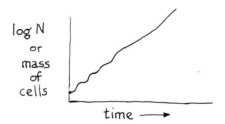

If the culture had started from a single cell, the progeny cells would divide more or less synchronously for a few generations, so that the number of cells would increase stepwise. Soon, however, the times of division become randomized and the number of cells in the culture then increases in a continuous fashion as does the total

mass **of** cells. The behavior of cancer cells or
other mammalian cells in culture is similar to
that of E. coli with the exception that liver
cells may grow 20 or 100 times more slowly than
E. coli.

Exponential growth   As long as the medium contains excess food for
the number of cells present, there is no competi-
tion for nutrients; each cell grows as if no
other cells are present, and the increase in
cells is proportional to the number of cells
present at a given time. Under these conditions,
the number N of cells (or the total mass of
cells) per unit volume in the culture at a given
time t is given by the equation

$$dN/dt = kN$$

(differential form of the growth equation)

or

$$N/N_0 = e^{kt} = 2^{rt} = 10^{\alpha t}$$

(integral form of above equation),[*]

where N_0 = number of cells at time t = 0 and α,
k,r = growth rate constants in various units;
specifically r is the <u>doubling rate</u> or number of
doublings per unit time. This is called exponential growth; a plot of log N versus time yields a
linear growth curve:

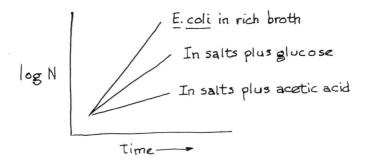

The growth rate of E. coli, that is, the slope
of the growth curve, varies depending on the
medium in which the cells are growing. In a rich

---

[*] $dN/dt = kN$; $dN/N = kdt$;

integrating: $\ln N = kt + constant$;

since $N = N_0$ when t = 0, $\ln N_0$ = constant, and so

$\ln N - \ln N_0 = kt$; $N/N_0 = e^{kt}$.

medium such as a good meat or yeast broth the doubling time is about 20 minutes. In a less rich medium containing nothing but salts, glucose for carbon, and an ammonium salt such as NH_4Cl for nitrogen the doubling time of the same bacterium is about 50 minutes. In a third medium containing an even less desirable food source, such as acetic acid instead of glucose, the bacteria double even more slowly (about every 120 minutes). [Why is acetic acid less good than glucose? Because more complicated and less efficient chemical pathways are needed to convert acetic acid to all the compounds needed by the organism.]

Chemical analysis reveals a remarkable fact: the chemical compositions of bacteria grown in different media are different. The reason for this is that under different conditions, different parts of the genetic program are expressed. In a rich broth, for example, most of the proteins whose function is to make the food substances already present in the broth become unnecessary. Hence, if one compares the proteins of bacteria that are growing in rich broth with those grown in a poor medium, different sets of proteins are found to be present.

If bacteria grown in a rich medium are switched to a poor medium, are the cellular components characteristic of bacteria grown in a rich or in a poor medium? At first, the protein components are those characteristic of the rich medium. But in the new medium new genes are turned on to make the extra enzymes (that is, the protein catalysts) needed for growth in the poorer medium. Thus there is a transition period during which the proportions of different proteins in the cell are changing. Eventually the cellular components will be exactly the same as for cells that had always grown in the poor medium.

Bacteria live in a changing environment and have been programmed by natural selection to respond to changes. The response of bacteria to environmental shifts illustrates the extreme precision with which the environment controls the function of bacterial genes. In contrast, liver cells or other human or mouse cells grown in culture do not respond very well to changes in the environment. Programmed for growing and func-

tioning well, protected from drastic environmental changes within the body, these cells have
lost the flexibility of response that bacteria
retain. Instead, in multicellular organisms,
different parts of the program are expressed in
different types of cells.

   The quantity of cells in a culture can be
measured in several ways:

   1. By counting cells with a microscope.
Bacteria are visible in the light microscope but
show little detail because their size is near its
resolving power.* Larger cells can be counted
electronically in a Coulter counter, which
registers a pulse every time an object passes
through a small orifice.

   2. By measuring the turbidity of the medium
with an optical instrument. When particles such
as bacteria, whose dimensions are of the same
order as the wave length of visible light, are
evenly suspended in a liquid, within certain
limits they scatter light in proportion to their
concentration. When a beam of light is passed
through such a suspension, the reduction in the
transmitted light gives a measure of the cell
concentration:

$$C = K \log (I/I_0),$$

where

I_0 = intensity of light entering the suspension
of bacteria,

I = intensity of light which emerges from the
suspension,

C = concentration of bacteria,

---

* The resolving power of an optical system is the
minimum distance d between two points for which
the two points can be distinguished. For a
microscope, this is

$$d = 0.6 \; \lambda/(n \sin \theta)$$

where n is the refractive index of the medium between the object Ob being viewed and the front
lens, θ is the half-angle between the object and
the lens, and λ is the wave length of light. The
limit of resolution of a light microscope is
about 0.2 μm. For an electron microscope it is
0.1 to 1 nm = 1 to 10 angstroms.

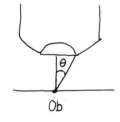

K = constant or "optical density" (for a suspension 1 cm thick).

A good spectrophotometer does the calculation "log I/I_o" and gives optical density readings directly.

3. By drying a measured portion of the cell culture and weighing the dried cell material (the dry weight of a bacterium is approximately 10^{-11} grams; that of a cancer cell may be 1,000 times greater). This method is rarely used because of its relative insensitivity: it is difficult to weigh accurately less than 1 mg, which represents the dry weight of a few billion bacteria or a few million liver cells.

In these ways one can measure the total quantity of cells, either living or dead, in a culture. The number of viable bacterial cells is determined by spreading a measured sample of a culture on one or more nutrient agar plates and incubating. The number of colonies that grow corresponds to the number of viable bacteria and is a linear function of the amount of culture that has been plated. That is, one colony can be produced by a single bacterium.

Similarly, one can plate out mammalian cells in culture and get colonies. But this is easier said than done. It is only within the last 20 years that cell biologists invented tricks and manipulations by which animal and human cells could be studied like microbes. The impact was so great that entirely new fields of biology were created. The microbes still have a lot to teach us.

Further reading

Watson's Chapter 3, "A Chemist Looks at the Bacterial Cell," makes an excellent transition between the material in this lecture and the ones that follow.

Pages 361B-370B in this book, "Chemical Background," are provided for the reader who needs to learn or refresh some basic elements of chemistry.

Lecture 2

Different kinds of cells in the human organism
grow and divide at different rates. Some, such
as nerve cells, do not divide at all after the
second year of life. Others, such as liver
cells, can still divide but have lost the ability
to make many essential molecules and in order to
grow in culture need complex media supplying all
the substances the cells cannot make for themselves. The same is true, of course, of the
whole human organism. Like all vertebrates, we
require many specialized nutrients. In contrast,
green plants and most of their cells can make
almost everything they need if supplied with
water, minerals in soil, and sunlight. Some bacteria are almost like plants in being able to
make everything they need; others are as fastidi-
ous as liver cells.

The growth curve for bacteria has an initial
phase (lag phase) in which cells grow more slowly
than expected because they have to adapt to the
new medium by making the proper enzymes. In the
final phase (stationary phase) growth stops,
generally because some nutrient is exhausted.
The nutrient that is depleted first is the <u>limit-
ing nutrient</u>.

If for example the ammonium salt NH_4Cl, which
provides nitrogen, is in a limiting amount, while
all other foods are in excess, the growth curve
looks like this:

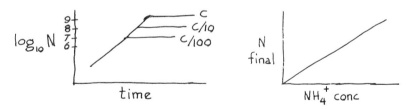

The bacteria continue to grow at a constant maximal rate as the concentration of ammonium salt in
the medium is constantly decreasing. The transition from the exponential phase to the final stationary phase is sharp and represents the deple-
tion of the nitrogen source. No growth occurs
after the nitrogen is exhausted because all of
the macromolecules of the cell contain nitrogen.

If we start several cultures with media differing only in the concentration of a single component, for example NH_4Cl, we get a set of growth
curves in which the final concentration N of

cells is proportional to the initial concentration of nitrogen in the medium: all nitrogen used up from the medium is found in cell material.

How does the doubling rate r (number of doublings per unit time) depend on the medium? The relation is (experimentally)

$$r_C = r_{max} \frac{C}{C + C_{1/2}} ,$$

where r_C is the growth rate constant in a medium with a concentration C of the rate-limiting nutrient and $C_{1/2}$ is the concentration that gives half maximal growth rate.

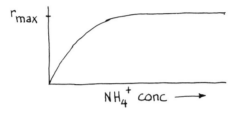

The dependence of growth rate on the concentration of a limiting nutrient is best studied in a special device, called a chemostat, which is a flow-through system that maintains constant conditions in the medium (see appendix at the end of this lecture). In a chemostat one can observe a simple but impressive example of evolution: a selection in which the population of existing bacteria is replaced, after several days, by a "better fit" population, that is, by a mutant that grows better or faster under the same conditions. Such a mutant, when it first appears, represents a very small minority, maybe 1 in 10^7 bacteria, but after many generations in the chemostat it will take over because it grows faster than the original cell type.

How could a mutant grow faster? Here are some suggestions from students:

1. It might make smaller cells. This makes more individuals but growth is not really faster: the total mass would not change.

2. It might develop a shortcut in a biosynthetic pathway. This is unlikely, because the biochemical pathways have been selected gradually over long periods of evolution.

3. It might conserve energy by moving less

and thus grow more. This solution is unlikely, although possible: motion energy is only a small part of the energy budget.

4. It might eat other bacteria.

5. It might have a more effective <u>pump</u> that concentrates an essential nutrient, for example glucose, and therefore might have an advantage over the other cells. This is what in fact occurs.

[Digression on control mechanisms:

1. Cell numbers can increase without increase in total mass. During the first few divisions following fertilization of a frog egg more cells are formed but there is no increase in total mass. Thus growth is not bound to cell division.

2. When a piece of the liver is removed from a rat, the remaining cells begin to grow and to divide until the original mass of the liver is restored. This indicates that cells that have been resting for many years retain the ability to grow if controls are removed. After removal of a tumorous kidney, the other kidney grows to compensate for the loss. We conclude that the normal size of an organ when growth stops is determined genetically through the interaction among various organs of the body.]

Cell composition

Because of the overall similarity of living matter, the chemical composition of all cells is basically the same. About 70 percent of the weight of bacteria and of most cells is water (determined by weighing cells before and after drying). The dry residue contains about 50 percent C, 10 percent N, 15 percent H, 20 percent O, and 5 percent other elements, including sulphur found in proteins, phosphorous in nucleic acids, and charged ions such as potassium or sodium.

Membranes

The amounts of various small molecules within cells are different from the amounts outside because of the presence of the cell membrane which has specific selective permeabilities. The concentration of some substances is maintained at a higher level inside the cell by specific pumps or <u>active transport</u> systems in the membrane. The membrane would be essentially impermeable to electrically charged atoms and molecules, that is, to <u>ions</u> were it not for the existence of specific transport systems for K^+, Mg^{++}, and amino acids. Similar transport systems also exist for electrically uncharged molecules like sugars and serve to transfer them from a region of lower concentra-

tion to one of higher concentration. The "backbone" of the membrane is a hydrophobic fatty
bilayer, which repels charged molecules because
these have great affinity for water.

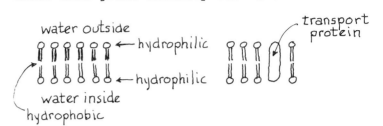

Since practically all substances produced as metabolic intermediates are charged, they do not pass
through the membrane and cannot leak out of the
cell. The passage of uncharged molecules depends
on their size and structure. For example, glucose enters a cell 1,000 times more slowly than
water.

A charged molecule is one that, when in water
solution, gains or loses one or more electrons,
usually in order to complete an outer electron
shell. (The shells have, starting from the inner-
most one, 2, 8, 8, ... electrons; see page 361B.)
For example,

H^+ has no electrons (1-1) and a positive charge;

Na^+ has 10 electrons (11-1) and a positive charge;

Cl^- has 18 electrons (17+1) and a negative charge.

Weak bonds

Water is a peculiar solvent. Because of the
electronegativity of the O atom the shared electrons are strongly attracted to the oxygen,

Lecture 2

and water is said to have a <u>dipole moment</u>. Any
negatively charged ion (or <u>electronegative</u> atom
in molecules, for example, :O: or N: is attracted
to the positive side of the water molecule to form
a <u>hydrogen bond</u>, in which the hydrogen is shared
by two electronegative atoms.

Hydrogen bond

This is a typical <u>weak interaction</u> with a bond
energy of about 3 kcal/mole (the bond energy is
the energy needed to undo the bond, for example,
by collision or thermal agítation). Other examples of weak bonds are the <u>ionic bond</u> between
charged ions, for example, Na^+-Cl^- in a salt
crystal, and the <u>hydrophobic bonds</u> between nonpolar (non-water-attracted) substances. Oil, for
example, contains long nonpolar fatty chains which
exclude water but interact with other nonpolar
substances.

Proteins, as we shall see, have polar groups
that attract water, and also nonpolar groups that
avoid it. The final conformation of a protein is
the result of a struggle between "those who like
the rain and those who don't." A mutation that
results in changing one polar for one nonpolar
amino acid can actually change the shape of an
entire protein molecule. Note that weak bonds
are not arbitrarily defined. They are those that
have a finite probability of being broken by
thermal motion, that is, by random collisions between molecules. The average energy transferred
in one such collision is kT = 600 cal/mol. The
rate of breaking of bonds with energy E is $e^{-E/kT}$.
(see page 387B).

Ions

Some water molecules are always dissociated in-
to ions: $H_2O \rightleftharpoons H^+ + OH^-$. Approximately 1 in 10^7
molecules of pure water are dissociated per mole
of solution. Taking the concentration of water
itself as unity, $[H^+] = [OH^-] = 10^{-7}$. and $\log_{10}[H^+]$
= -7. By definition, pH = $- \log_{10}[H^+]$. For pure
water, pH = 7. Addition of some NaCl to water
(NaCl $\rightleftharpoons Na^+ + Cl^-$) does not change the pH because
it adds equal numbers of + and - ions and does not
cause release or uptake of H^+. If instead one
adds hydrochloric acid HCl, the pH drops, since
the HCl ionizes (HCl $\rightleftharpoons H^+ + Cl^-$) releasing excess
H^+ ions. NaOH added to water releases OH^- ions
(NaOH $\rightleftharpoons Na^+ + OH^-$), which raise the pH by removing
H^+ from solution: the reaction $H_2O \rightleftharpoons H^+ + OH^-$ is
pushed toward formation of water by the increase
in OH^- concentration.

Acids and bases

An <u>acid</u> (for example, HCl) is a substance that, when added to water, releases H^+ and therefore lowers the pH; a <u>base</u> (for example, NaOH) is a substance that, when added to water, releases OH^- or takes up H^+ and therefore raises the pH. For example, consider ammonia NH_3 in water:

$$
\begin{array}{c}
H \\
\cdot\cdot \\
H \!:\! N \!:\! \\
\cdot\cdot \\
H
\end{array}
\underset{-H^+}{\overset{+H^+}{\rightleftharpoons}}
\left(
\begin{array}{c}
H \\
\cdot\cdot \\
H \!:\! N \!:\! H \\
\cdot\cdot \\
H
\end{array}
\right)^+
$$

The net result of the reaction $NH_3 + H^+ \rightleftharpoons NH_4^+$ is a decrease in H^+ in the solution and therefore a relative increase in OH^- and an increase in pH. Thus ammonia is a base; NH_4^+ is an acid. A pair of substances like NH_3 and NH_4^+ is referred to as <u>conjugate base</u> and <u>conjugate acid</u> (see page 366B). (Note that the positive charge in the NH_4 ion is not localized on any one atom: it is shared.) NH_4^+ is called the ammonium ion.

One final comment: an organic acid, for example, pyruvic acid, $CH_3-CO-COOH$, dissociates into $CH_3-CO-COO^-$ and H^+. Most physiological solutions, such as blood or cell sap, have pH near 7. At this pH most organic acids are fully dissociated, as we shall see later. They may, for convenience, be considered to be in a salt form because of the great excess of Na^+, K^+, etc. in the solution. (A <u>salt</u> is a compound in which a hydrogen atom of an acid has been replaced by a metallic atom.) Hence, biochemists often use the terms pyruvic acid and pyruvate, or succinic acid and succinate, interchangeably and so shall we in the rest of this book.

Covalent bonds

We have discussed a variety of weak bonds and their role in molecular conformation. Within most molecules the atoms are held by <u>covalent</u> <u>bonds</u>. Covalent bonds join atoms together in molecules such as water H_2O, ammonia NH_3, and carbon dioxide CO_2. They are stronger and tighter than the weak bonds we have discussed; they require an energy of about 100 kcal/mole to break, that is, to pull the atoms apart. Most reactions of biological interest do not involve breakage of covalent bonds but replacement of atoms and groups of atoms.

Covalent bonds have shared electrons. In a

single bond (:) there are 2 shared electrons; in
double bonds (::), 4 shared electrons; in triple
bonds (:::), 6 shared electrons. Single-bonded
atoms are farther from each other than double-
bonded atoms, etc.; for example, the distance in
C:C bonds is = 1.54 A; in C::C, 1.39 A; in C:::C,
1.20 A. Thus the bonds not only determine the
strengths with which the atoms are held together
in a molecule, but also their distances apart and
therefore, they determine the molecular structure.

In single bonds, but not in double and triple
bonds, the atoms are free to rotate around the
axis of the bond as a steering wheel can rotate
around its axis, except for steric constraints.
What are steric constraints? Suppose, for example,
two adjacent atoms joined by a single bond have
attached to them groups of other atoms. When the
two atoms rotate around their single bond the
attached groups may bump into one another; that
is, they may come so close that their electronic
domains begin to overlap and repel each other.
Thus, although in theory the atoms can rotate
around the single bond joining them, they cannot
always do so because of the repulsions of the
side groups.

Double bonds and triple bonds are rigid, that
is, they permit practically no rotation. All
this information will be useful to us later on in
understanding the structure of complex molecules.

Before we come to that, a digression. Let us
ask, What are cells composed of?

Chemical
fractionation

Suppose you have a suspension of 1 gram of
bacteria (or liver cells) in water. You centrifuge it and resuspend the cells in a little bit
of water. Now add acid so that the concentration
of H^+ is about 1/10 M (pH = 1). The small
molecules in the cell generally stay in solution
since they are almost all electrically charged
and can compete with H^+ ions for water. They are
acid soluble. The large molecules such as proteins and nucleic acids are held in solution
primarily by weak interactions within themselves
and with water. The acid destroys these weak
interactions by taking away water. They are,
therefore, acid insoluble. About 20 percent of
the dry weight of the cell is acid soluble, the
rest is acid insoluble.

The acid-soluble fraction includes all small
molecules: some are building blocks of larger

Cell lipids

Phospholipids

molecules; some are products of degradation of
other large molecules that may again be reutilized;
and certain others are part of the chemical
machinery of the cell. The acid insoluble fraction is composed of larger molecules of which
there are four classes: proteins, nucleic acids,
polysaccharides, and lipids (fats), all of which
can be separated by appropriate fractionation.

To separate rather brutally the various classes
of large molecules you might begin, for example,
by extracting the lipids by shaking the acid insoluble fraction with a nonpolar solvent such as
chloroform plus methanol. Most of the fats
have chains of methylene groups $-(CH_2)_n-$ ending
in a methyl group $-CH_3$, all nonpolar groups which
do not like water and readily pass into nonpolar
solvents. The extracted lipids can then be
analyzed in many ways.

What functions do fats serve in the cell? Some
are reserve fats. These are mostly found in
special fat-storing cells, which also contain a
lot of water. Fat people are heavy because of the
water they store in expanded fat tissue rather than
because of the weight of the fat itself.

Another class of lipids are phospholipids, the
essential components of all cellular membranes.
They consist of two nonpolar chains of fatty
material attached to a molecule of glycerol. Attached to the third position of glycerol is a
phosphate group which is in turn attached to any
one of several other compounds.

CH_2OH CH_2-O- fatty acid 1 (straight or bent)

$CHOH$ $CH-O-$ fatty acid 2 ("   "   ")

CH_2OH $CH_2-O-\overset{\overset{O}{\|}}{P}-O-CH_2CH_2NH_2$

glycerol O^-

The phosphate group in phospholipids likes water.
To make membranes evolution has brought into
being a molecule with a head that likes water and
two legs that do not like it. These molecules
arrange themselves like this: the charged surfaces are directed toward the aqueous environments
inside and outside the cell. The nonpolar groups in
between form a bilayer that prevents the passage

of many substances except when there are specific transport mechanisms or pores (see illustration on page 18).

Of course, cellular membranes do not consist of phospholipids only. Embedded within the bilayer there are all sorts of proteins, which are the active components of the membrane. Cell membranes have some truly fantastic properties, at least in bacteria. One of them is the ability to regulate their own composition so as to maintain a more or less constant fluidity. The degree of fluidity, that is, the softness or hardness of the membrane, depends on the nature of the fatty acids in the phospholipids: the more straight-chain, saturated fatty acids there are, the more rigid the membrane, and vice versa for the bent-chain, unsaturated fatty acids. The reason is simple: straight chains pack more tightly, bent chains tend to disrupt the packing. The lower the temperature, the tighter the packing. To maintain the same fluidity, cells growing at lower temperatures incorporate more bent-chain fatty acids, and vice versa.

How can they do that? Nobody knows for sure. It is clear that there must be a sensor, which indicates the temperature or some correlate of it, and an effector that responds to the signal. The sensor must function as a thermometer, while the effector in some way must regulate the amount of unsaturated fatty acids that are put into the phospholipids as they are synthesized and inserted into the membrane. Sensor and effector could both be properties of the enzymes that make the phospholipid molecules.

Polysaccharides

The next class of large acid-insoluble molecules consists of the polysaccharides. They consist of chains of simple sugars (monosaccharides). The chains may or may not be branched.

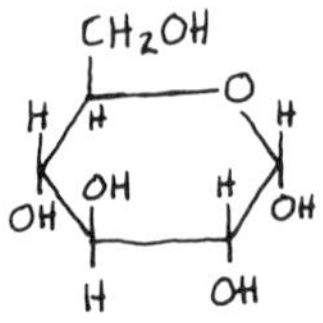

glucose

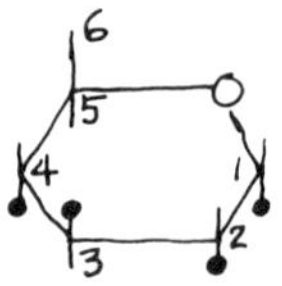

abbreviated
representation
and numbering
of C atoms

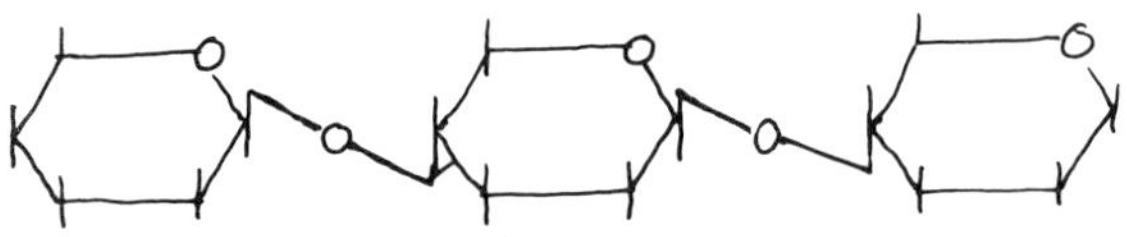

a polysaccharide
(one H_2O eliminated at every addition)

For every two sugars joined in a polysaccharide, one molecule of water is removed. This is the way all polymeric molecules are made, as we shall soon see. Polysaccharides have been developed in the course of evolution in many ways and for different uses. Some are reserve materials. For example, glycogen, a branched chain polymer of glucose, is the storage form of sugar in our muscles. When we exercise moderately, the source of energy for muscle contraction is not glucose from the blood but glycogen that has been stored in the muscles and is therefore directly available. These reserve materials are not made only in exceptional circumstances but are part and parcel of the functioning apparatus of cells; yet their amounts vary from cell to cell and from time to time, depending on the functional state (for example, rest versus exercise).

Polysaccharides also serve some structural functions by acting as skeletal material. Our internal skeleton is made of bones; the skeletons of some fishes are made of cartilage. Both bone and cartilage are not polysaccharides; they are essentially made of a protein called collagen impregnated with various salts. The situation is different, however, in bacteria, molds, yeasts, plants, and arthropods (insects, spiders, lobsters) all organisms that have their skeletons on the outside (exoskeletons). The outside skeletons

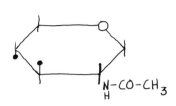

$N-CO-CH_3$

consist primarily of polysaccharides. The skeleton of plants is made of cellulose, a polymer of
glucose; the skeleton of insects and crustacea
is made of chitin, a polymer of a more complex
sugar called N-acetylglucosamine. In all these
exoskeletons, the sugars are linked together in a
special way which, according to organic chemists,
is the best possible way to make chains that can
lie in flat bundles and layers side by side. Once
again we see that natural selection knows its
business.

Appendix:
The chemostat

The Chemostat

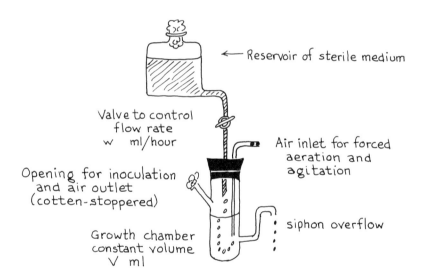

In the chemostat nutrient medium flows into a
growth chamber, whose contents are mixed by a
stream of filtered air. The excess fluid exits
through the siphon overflow and can be sampled for
bacterial counts and other measurements.

When the chemostat operates under steady-state
conditions, the number N of bacteria per unit
volume does not change: as many bacteria are made
as are washed out with the effluent. The growth
rate is limited by the concentration c of a limiting nutrient in the growth chamber. When $k_c = w/V$,
$dN/dt = 0$, and the bacterial concentration remains
constant at a value N determined by $a = (A-B)/N$,
where

A = concentration of limiting nutrient per milliliter in the reservoir;

B = concentration of limiting nutrient per milliliter in the effluent in the steady-state condition.

When the chemostat is operating at very high efficiency, $B \approx 0$; and $a \approx A/N$. The change per unit time is

$$dN = k_c N\, dt - \frac{w}{V} N\, dt,$$

$$\frac{dN}{dt} = N(k_c - \frac{w}{V}),$$

where

k_c = growth rate constant,

w = flow rate in milliliters per unit time, and

V = volume of the growth chamber in milliliters. (Note that w/V is the "washout rate," that is, the proportion of the contents of the growth chamber removed per unit time.)

Mutations in the chemostat

Assume that mutant bacteria are produced at a constant rate from normal bacteria. For mutants that grow at the same rate k_c as the normal bacteria, then

$$\frac{dM}{dt} = k_c M - \frac{w}{V}M + \mu N = \mu N$$

where the mutation rate is μ and the number of mutants is M. The number of mutants increases linearly. For certain mutants, however, the growth rate constant may be greater or lower than that of the parent cells! For a mutant that grows faster, its cells are washed out more slowly and accumulate until they represent the majority of the population. This can happen over and over (periodic selection) and provides an elegant demonstration of evolution by mutation and selection in a constant environment.

Further reading

Watson's Chapter 4, "The Importance of Weak Chemical Interactions," is packed but rewarding.
Lehninger's Chapter 10, "Lipids, Lipoproteins, and Membranes," supplements material in this and the next lectures.

Lecture 3

Even though polysaccharides can be clever, in the sense that they are properly structured for the jobs they have to do, as we have seen in the preceding lecture they are really dull molecules, repetitious and with little leeway for variation and imagination. They certainly do not have the characteristics one would want for the program molecules.

Proteins and nucleic acids

Thus we come to the other two classes of acid insoluble materials: proteins (about 50 percent of the dry weight of the cell) and nucleic acids (about 15 percent). As they are present in our acid precipitate, these substances are still intact as molecular entities, but are no longer in their natural state. They are still held together by their covalent bonds, but many weak bonds are broken. We say that these molecules have been denatured.

As we take a closer look at the protein and nucleic acid fractions from our gram of cells, we must keep in mind that they are mixtures of many kinds of proteins or nucleic acids. In order to isolate just one species, for example, a single enzyme, one would have to proceed differently: break the cells gently; keep the extracts cold all the time; fractionate them in an ultracentrifuge (a centrifuge that generates enough centrifugal force to sediment large molecules); separate the components on columns of chromatographic resins; and use many other techniques of biochemistry. In this way one can get a pure protein or nucleic acid with greater or lesser ease. For example, it is easy to purify hemoglobin from red blood cells just by diluting blood in water: the red cells burst and release their contents, 95 percent of which is hemoglobin (about 10^6 molecules per cell). When one wants to isolate a rare enzyme of which there may be only 100 molecules per cell, the problem becomes much more difficult.

Proteins

Independently of their source, all proteins have a major property in common. If they are heated with hot concentrated acid, they break down into units called amino acids; the strong acid hydrolyzes the bonds that join the amino acid in the protein molecules; that is, adds one molecule of water for every linkage it breaks.

Amino acids

All proteins consist of the same 20 amino acids in different molar proportions. Refined analysis shows that these amino acids form unbranched

chains, the <u>polypeptide</u> chains (a fragment of protein is called a <u>peptide</u>, and the linkage between
amino acids is a <u>peptide</u> bond; more on that
shortly). The amino acids found in proteins have
four groups bonded to a central carbon atom
called the α-carbon: (1) a carboxyl ($-\overset{O}{\overset{\|}{C}}-OH$,
abbreviated -COOH), (2) an amino group H-N-H or
$-NH_2$), (3) a hydrogen atom, and (4) a side chain
or R group, which is different for each amino
acid (successive carbons of the R group, if there
are any, are designated β,γ...):

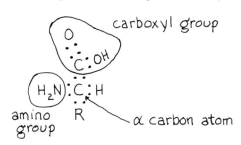

In all of the amino acids except glycine (R=H)
the α-carbon is <u>asymmetric</u> because it has four
different groups attached to it. Molecules with
asymmetric C atoms rotate the plane of polarization of polarized light--a useful property for
analytical purposes.
    More important is the fact that each amino
acid has at least two groups that can release or
take up H^+ ions:

the carboxyl group: $-COOH \rightleftharpoons -COO^- + H^+$

the amino group: $-NH_2 + H^+ \rightleftharpoons N^+H_3$

These groups have different tendencies to take up
or give up hydrogen ions; these tendencies are
measured by the equilibrium constants of the
reactions. At equilibrium, the forward reaction
balances the backward reaction. The <u>equilibrium
constant</u> is the ratio of concentrations of the
products divided by the concentration of the reactants.

$$K_{a\ eq} = \frac{[COO^-][H^+]}{[COOH]} \quad \begin{array}{l} products \\ reactants \end{array}$$

Taking the logarithm of both sides of the equation
we get

$$-\log_{10}K_{a\ eq} = -\log_{10}[H^+] - \log_{10}\frac{[COO^-]}{[COOH]} ,$$

$$pK_a = pH - \log_{10}\frac{[base]}{[acid]} .$$

pH and pK_a

Each group that can release H^+ has a characteristic tendency to dissociate, measurable by the constant pK_a. By changing the pH, that is, the concentration of H^+ ions, one can cause a carboxyl group (for example) to pass from primarily -COOH (the acid) to primarily $-COO^-$ (the conjugate base) (see page 366B).

The carboxyl group is an acid because it has a tendency to give off H^+ ions; conversely, the amino group $-NH_2$ is a base because it tends to pick up H^+ ions. $-N^+H_3$ is the <u>conjugate acid</u> of the base group $-NH_2$; $-\check{C}OO^-$ is the <u>conjugate base</u> of the acid group -COOH. Even water H_2O is a base because it tends to attract H^+ ions: $H_2O + H^+ \rightleftharpoons H_3^+O$. The greater the tendency of a base to pick up H^+ ions, the stronger it is (H_2O is a very weak base). Conversely, the stronger its tendency to give up H^+ ions, the stronger an acid is.

Because of its ionizable groups, the form and charge of an amino acid change when pH changes.

at low pH, charge +1   at medium pH,   at high pH,
                        charge 0   charge −1

The pK_a values for carboxyl groups in different amino acids, although close, are not identical. This is because the pull on the electrons surrounding the carbons varies depending on the nature of the R group attached. The stronger the pull electrons receive from other parts of the molecules, the lower the pK_a. [Can you see why?] The more similar the R groups of two amino acids, the closer their pK_a values are (see page 367B). In addition to the -COOH and $-NH_2$ groups attached to the α carbon atom, some of the amino acids have in their side chains other acid/base groups, which also ionize and can contribute to the total charge of the molecule.

Electrophoresis

Because of these differences in the nature of their R groups, the amino acids can be separated by electrophoresis. In this process the substances to be separated are allowed to migrate to one or the other pole in an electric field created by an applied potential. The speed at which they migrate depends on the size of the molecules and on the charge of the molecules. For amino acids, if the side chains contain no charged groups, migration rate depends on the slight differences in the charges of the α-amino and α-carboxyl groups at different pH's.

(Note: Charges are not always integral, that is, +1, -2, etc. Fractional charges can occur by having some of the molecules with one charge and some with another. When the pH is near the pK_a for one of the dissociable groups, that group is dissociated on only some of the molecules. A charge of 0.3, for example, means that 30 percent of the molecules have a charge of +1 and 70 percent have a charge of 0. This does not mean that for every 100 molecules in solution 30 have a charge of +1 and maintain this charge. Rather, protons are continually going back and forth on and off the groups that can accept them, so that on the average 70 percent have charge 0, 30 percent have charge +1.)

Electrophoresis can be done in a variety of ways on a variety of substances. In filter paper electrophoresis, a mixture of amino acids is spotted on paper moistened with a salt solution, with 2 electrodes immersed at the two ends, and a potential difference is applied:

The direction and velocity of motion of a particular substance on the paper depends on its charge at the pH chosen for electrophoresis (pH of the salt solution):

$$V = qE/f_0 ,$$

where V = velocity, q = charge, E = potential difference, f_0 = frictional coefficient (which

depends on the size and shape of the molecule).
Thus a small difference between the charges of
two substances results in their migrating at different velocities. When possible, the pH is
chosen to maximize the minute difference between
the molecules to be separated.

Chromatography

Amino acids can also be separated by chromatography on paper (or on other suitable carriers).
The mixture is spotted on paper and put in a
closed vessel with a mixture of H_2O and a nonpolar solvent (for example benzene). Water vapor
moistens the paper. As the nonpolar solvent
starts creeping on the surface of the paper, carrying along various substances from the spot, each
substance moves to a distance proportional to the
ratio of the time it spends in the nonpolar solvent to the time it spends in water. Thus chromatography exploits differences in hydrophilic and
hydrophobic behavior: a more polar substance
spends more time in the water and moves less far.
An amino acid with a nonpolar R group on its side
chain (for example, alanine R = CH_2CH_3) does,
therefore, move farther than an amino acid with a
polar R group (for example, serine R = CH_2CH_2OH).

Peptide bond

Proteins consist of polypeptide chains: long
chains of amino acids in specific sequences,
attached together by peptide bonds, in which two
amino acids are joined with elimination of water:

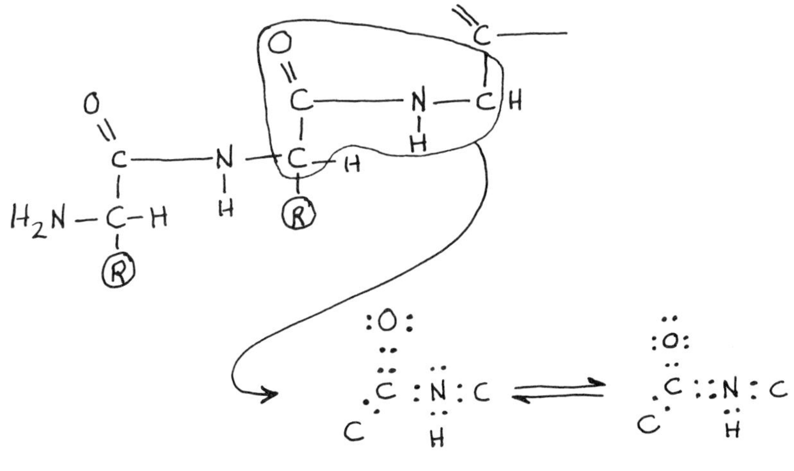

peptide bond, planar

(It is a general rule that biological polymers are
chains of monomers minus one molecule of water for

each link.) The peptide bond is <u>resonance-sta-</u>
<u>bilized</u> by <u>delocalization</u> of the lone pair of
electrons of the N atom: The importance of this
is that almost no rotation is possible around the
peptide bond because it has a partial double-bond
character. This limits the mobility of the chain
by placing a constraint to rotation and plays a
major role in determining the overall structure
of a protein. The six atoms C(CO)(NH)C are in
one plane. Rotation is possible about the other
bonds of polypeptide chains, except where it is
prevented by steric hindrance between bulky side
groups.

   Each polypeptide has an <u>amino end</u>, with a free
amino group, and a <u>carboxyl end</u> with a free carboxyl group. Notice that in making a polypeptide
the charged α-amino and α-carboxyl groups are
used to form peptide bonds. Thus the only charged
groups remaining on the backbone of the polypeptide are the terminal amino group and the terminal
carboxyl group.

$$H_2N - \underset{\underset{H}{|}}{\overset{\overset{R}{|}}{C}} - CO-NH ------ NH - \underset{\underset{H}{|}}{\overset{\overset{R}{|}}{C}} - COOH$$

amino terminal   carboxyl terminal

The backbone, therefore, is informationally
uninteresting. What gives proteins their individuality, their meaning, is the order and properties of the side groups attached to their
utterly boring backbone.

Sequences

   The only thing that "makes sense" in protein
molecules, that is, carries information, is the
sequence of side groups. How do we learn this
sequence?

   This first step in sequencing is to get a preparation of a pure protein, in which all molecules
are alike. Then one can apply several techniques.
For example, the protein may be split specifically
by using a reagent that breaks the chain at certain chemically defined points. The enzyme trypsin breaks polypeptides on the -CO- side of the
amino acids lysine and arginine. Other reagents
split the chain at different points. Then the
various pieces can be separated and sequenced
individually (a task easier than sequencing the

entire chain). Finally, the order of the pieces
is determined by overlap comparisons:

Reagent I   _____ ____ _____ ___

Reagent II   ___ _____ _____ ___ ____

Peptides

The outcome of many analyses of this kind is
that each protein consists of a unique sequence
of amino acids, in one or several polypeptide
chains. Some proteins have one chain only, others
have 2 or 3 identical chains, others have chains
of two or three or four kinds. Also, each chain
may have attached to it some polysaccharide or
lipid chain or some other functional group. For
example, myoglobin, the red protein of muscle,
has a single chain of 153 amino acids with a heme
group attached to it. The heme is a flat molecule
which, in myoglobin and hemoglobin, serves to
store and release oxygen molecules when they are
needed. The O_2 molecule combines with the iron
atom in the center of the heme. Hemoglobin, the
red protein of red blood cells, has four chains
similar but not identical to the one in myoglobin,
each with a heme. The four chains are not identical among themselves, however: there are two α
and β chains, so that the formula of hemoglobin
would be $\alpha_2\beta_2$.

Taken by itself, the amino acid sequence of
any protein still appears to be meaningless. That
is, there is no regularity of grouping and no
obvious rule of neighborhood between amino acids.

The meaning of a protein, in two different
senses, comes from its use and from its history.
From its use, because the chain or chains of
amino acids, nonsensical as they seem on paper,
acquire chemical meaning by a process of "balling
up" of the chains into precise three-dimensional
structures, within which chemical surfaces are
created that have, for example, specific enzyme
activities. From their origin as gene translates,
the amino acid sequences acquire the meaning of
having been perfected by evolution for satisfac-
tory function in a given organism.

After the order of the amino acids in the
polypeptide chains of a protein is determined by

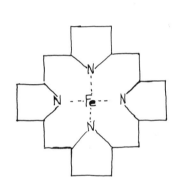

Protein structure

sequencing, the three-dimensional structure can be determined by x-ray diffraction: a beam of soft x rays passing through a crystal of a substance yields a pattern of reflections that can be used to derive the position in space of the various atoms and groups of atoms. Myoglobin was the first protein for which the position in space of each of its 153 amino acids was established by x-ray diffraction.

How does the three-dimensional structure of a protein arise? Fully by the interactions between the amino acids of its polypeptide chain or chains. These interactions are mainly weak interactions, that is, weak bonds that one-by-one can easily be broken--but when they are many, their strength holds the molecule firmly together and in a definite (although flexible) shape. You already have heard of some of these bonds:

1. hydrogen bonds, for example, between O and N atoms of neighboring amino acids;
2. electrostatic bonds between charged groups such as $-COO^-$ and $-N^+H_3$;
3. Hydrophobic bonds, such as the attraction between two or more chains $CH_3-CH_2-CH_2-$ which tend to escape from water;
4. disulfide bonds S-S between two -SH groups belonging to cysteine amino acids in different parts of the same chain or in different chains (-S-S- bonds are not as weak as the others and play a big role in generating major constraints in the shape of several proteins).

All of these bonds are not made by catalysts. They form spontaneously and, as they form, the molecules progressively tend to a state of minimal free energy, a more stable form. If a small polypeptide molecule is denatured, so that all its weak bonds are broken, and then allowed to renature by itself, it often regains its precise original structure. Thus this structure must depend only on the intrinsic nature of the molecule, that is, its amino acid sequence.

With this in mind, we say that a protein has four types of structures: (1) primary structure-- the sequence of amino acids; (2) secondary structure--the conformational changes in primary structure due to the formation of electrostatic and hydrogen bonds between nearby amino acids, for example, the so-called α-helix formed by many stretches of protein chains; (3) tertiary struc-

ture--the ultimate configuration that a polypeptide chain takes in reaching the configuration of
minimal free energy by folding and forming the
maximum possible number of weak bonds. The
tertiary structure of a protein tends to place
hydrophobic groups, methyl ($-CH_3$) and methylene
($-CH_2-$) groups, on the inside and the hydrophilic
groups on the outside; (4) quaternary structure--
association of polypeptide chains in proteins that
have more than one. In hemoglobin for example,
the two α and the two β chains have primary,
secondary, and tertiary structures; $\alpha_2\beta_2$, the
whole molecule, has a quaternary structure.

Active site

Once it has assumed a final configuration a
protein becomes functional. The lysozyme molecule,
for example, has two articulated chunks that
define a groove. In the groove is the active site
of the enzyme, which splits the polysaccharide of
the wall of bacteria at a specific place. Two
amino acids on the two sides of the groove project into the groove their side chain carboxyl
groups, which alter the electronic configuration
of the bond to be split and make it vulnerable
to splitting by a molecule of water. When in 1967

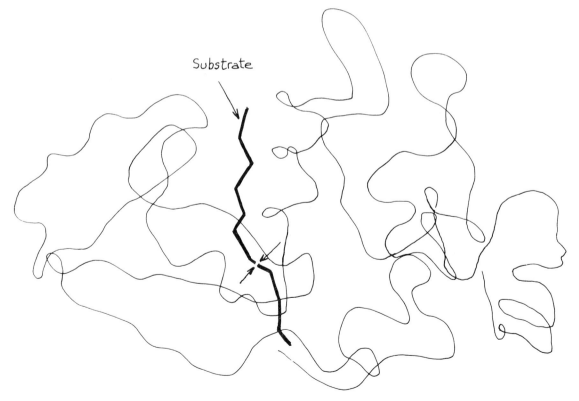

Substrate

the structure of the lysozyme was solved it was a pleasant surprise to the biochemists that the structure also explained the chemical function of the enzyme. It was the first instance in which the action of an enzyme was understood in visual terms at the level of the electronic distortions of the substrate, of the protein groups involved, and of the actual positions of the substrate and the active site of the enzyme molecule.

Since a protein molecule has a definite shape, is it a solid? Yes and no. The real question is, what is a solid? In a crystal such as NaCl, the ions Na^+ and Cl^- are in fixed positions. In a protein molecule, the atoms are in relatively fixed positions but subject to distortions. The peptide bonds are quite rigid, but all the other covalent bonds of the backbone can rotate. The weak bonds can open and close. When a substrate approaches an enzyme and attaches to its active site, the enzyme changes shape. When substrate or products come off, the enzyme relaxes and resumes its original shape. These changes are known as induced fit. Thus we must think of protein molecules not as hard rigid objects, but as slightly soft and able to modify their structure in response to conditions. As we shall discuss later, not only the substrate molecules, but also regulatory substances can change the shape of enzymes by attaching to them at various sites.

How big are protein molecules? One can get information about the size of a protein before knowing its specific structure by[*]-
using filters: but the values are accurate only within a factor of 3;
electron microscopy: but this is generally too coarse for a small protein molecule (4 nm in linear dimension);
mass spectrography: difficult to use with molecules of this size;
using an ultracentrifuge: the best way. In an ultracentrifuge substances move in the same way as in a gravitational field, with the centrifugal force replacing gravity. The motion of particles in solution in a centrifugal field is determined by the specific gravity (mass/volume ratio) of the particles, their shapes (which influence the frictional drag), and the strength of the field.

---

[*] Student's suggestions.

The motion of particles in water is defined by
Stokes' law:

$$\text{velocity} = (\omega^2 r) V (\rho - \rho_0)/f_0 \; ,$$

where $\omega^2 r$ = centrifugal acceleration, V = volume
of the particle, ρ = density of particle, ρ_0 =
density of the medium, and f_0 = friction coefficient. For a sphere, $f_0 = 6\pi R\eta$, where R = radius
of the particle and η = viscosity of the medium.
For nonspherical particles there are ways to
calculate the axial ratio, which measures how
different the molecule is in shape from a sphere.
Some proteins are almost spherical, others
resemble fat cigars or long rods.

A few proteins, such as silk, are filamentous,
with no balled-up tertiary structure. By far the
most important class of filamentous macromolecules
in living cells is the nucleic acids, which will
be the subject of the next lecture.

Further reading   Pages 361B-379B will be helpful at this
point.

Lecture 4

We now pass from proteins to nucleic acids. Let us do so by the way of sugars. A typical sugar molecule has one of the two forms:

$$CH_2OH$$
$$HC-OH$$
$$HC-OH$$
$$HC-OH$$
$$CHO$$

Ribose

A 5-carbon sugar molecule can be any one of several sugars depending on the configuration at each carbon, that is, depending on which side of the carbon the H and OH groups are. Let us take the 5-carbon sugar ribose. The -CHO group, called

aldehyde, has the structure $-\overset{\overset{O}{\|}}{C}-H$. This is dif-

ferent from the carboxyl group $-\overset{\overset{O}{\|}}{C}-O-H$. The carboxyl group is quite stable because of <u>resonance</u> (as in the peptide bond), but the aldehyde group is unstable and very reactive. [This is because the carbon atom is less <u>electronegative</u> than the oxygen atoms, so that electrons are pulled away from the carbon and toward the oxygen, and the carbon, deficient in electrons, becomes a target for all sorts of nucleophilic atoms, that is, atoms eager to share electrons.] Because of the reactivity of its aldehyde group, a sugar like ribose spends most of the time in solution in closed ring form. This form is called a furanose ring and results from a reaction between the hydroxyl group of carbon 4 and the aldehyde. The furanose structure of the 5-carbon sugar is part of the backbone of all nucleic acids.

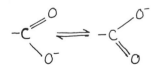

There are two categories of nucleic acids: those whose backbone contains ribose (ribonucleic acid, that is, RNA) and those whose backbone contains deoxyribose (deoxyribonucleic acid, that is, DNA) (see page 383B):

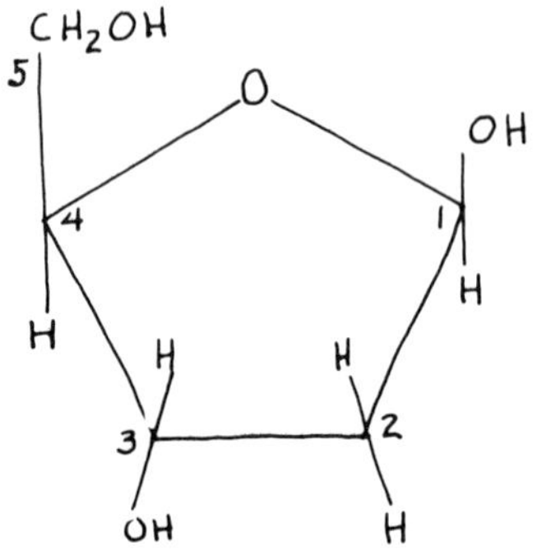

deoxyribose

In the nucleic acids, the sugars are joined together by phosphodiester bonds. (An ester is
the bond between an acid -COOH or OP(OH)$_3$ and an
alcohol -CHOH by removal of water.)

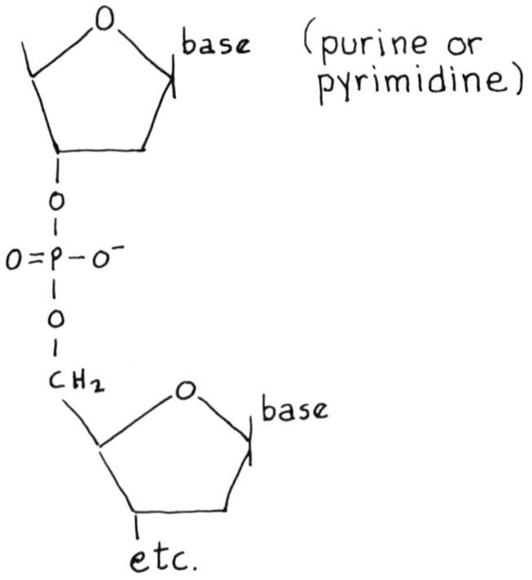

base (purine or
       pyrimidine)

base

etc.

The phosphodiester bonds create a P-sugar-P-
sugar...backbone, where P stands for phosphoric

acid (O-P-O). The sugars themselves are fairly

rigid because they are closed rings; rotation,
however, is possible about the phosphodiester
bonds.

Nucleotides

     The nucleic acid subunit or nucleotide consists
of a sugar-phosphate with a purine or pyrimidine
base attached to the sugar:

pyrimidine
(uracil)

nucleoside
(riboside)

(ribose)

HOCH₂

OH   OH

base   ——————

nucleoside ——————+ sugar

nucleotide ——————+——————+ phosphate

purine
(adenine)

NH₃

HC

to
ribose

Four bases are found in DNA: adenine (A), guanine
(G), cytosine (C), and thymine (T). Adenine,
guanine, and cytosine are also found in RNA, but
here uracil is present instead of thymine. A
nucleotide without its phosphate is a <u>nucleoside</u>.

Thus RNA differs from DNA in two chemical ways:
it has ribose instead of deoxyribose and uracil
instead of thymine. Note that the backbone of
each type of nucleic acid is the same throughout
all molecules in all living species. All the
information, therefore, is in the sequence of the
bases. Compare this with proteins: there too
the backbone is always the same, and all information is in the sequence of amino acid side chains.

DNA is a double helix, the celebrated double
helix of Watson and Crick: two polynucleotide
chains are held together by hydrogen bonds between
adenine and thymine (two bonds, A=T) and between
guanine and cytosine (three bonds, G≡C) in the two
chains. DNA in nature is almost always in the
double-stranded helical form, held together by
precise sequential base pairing between <u>comple-</u>
<u>mentary</u> sequences of nucleotides, like AAGATAC and
TTCTATG.

$$\left(\begin{array}{c} A=T \\ C\equiv G \end{array}\right)$$

Complementarity   The structure of DNA, as derived from x-ray

diffraction, is very precise. This plays an important role in the stability and reproduction of
DNA, which as you already know, is the essence of
the genes. Single-stranded nucleic acid chains
do not have a precise structure but form "random
coils." Occasionally, even in single-stranded
nucleic acids base pairing produces loops or
hairpin structures.

hairpin structure
in RNA

random coil, with

base-paired regions

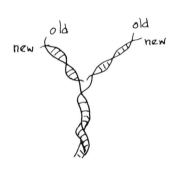
old   old
new   new

This happens, for example, to RNA molecules,
which are mostly single stranded. But the base
pairing that produces loops or hairpins is mostly
(not always) accidental and meaningless. Instead, the pairing between the two strands of
DNA, as well as the pairing between a strand of
DNA and the strand of RNA that is made upon its
direction, is very meaningful. In fact, all
usable transfer of information in biological
material--from viruses, to bacteria, to plants,
to man--is made through the specific pairing of
complementary sequences of bases. When DNA is
reproduced, the helix opens up and each strand
serves as template for the synthesis of a new
strand according to the rules of complementary
pairing A=T and G=C. One double-stranded DNA
molecule generates two DNA molecules, each of
which is identical to the original one. We shall
defer discussing the evidence for these statements
until we come to DNA and RNA synthesis.
     In general (except in the case of certain
viruses) RNA is made from DNA. The DNA in a certain region of a double helical molecule unwinds
so that the mechanism for making RNA can get
access to it. Only one of the two DNA strands is
transcribed by the usual rules of pairing, except
that in making RNA each adenine in DNA serves to
put uracil in RNA.

CTTCA   DNA
CUUCA
RNA   GAAGT

Even in the synthesis of a protein under the direction of an RNA template, as we shall see later in greater detail, information is not transferred directly from nucleic acid to amino acids. What is used to insert an amino acid into the growing protein chain is an amino acid attached to a specific piece of RNA called transfer RNA; the actual recognition is done by base pairing between the template RNA and the transfer RNA, which serves as "adaptor" to position the right amino acid where the template directs it. The transfer RNA has a specific 3-dimensional structure resulting from internal base pairing sequen sequences.

Transforming principle

The first solid evidence that DNA is the material of the genes came in 1943. It was known that if one took certain mutant bacteria and mixed them with an extract of normal bacterial cells of the same species, some cells became like those that had provided the extract and gave rise to transformed progeny cells. Oswald Avery and his co-workers showed that the active component of the extract was DNA: the more they purified the DNA the higher the frequency of transformations they observed. Any one of the hereditary characteristics of the bacteria they were studying could be transformed with extracts from some other bacterial strain. The bacteria studied by Avery were pneumococci, that is, pneumonia bacilli, but similar results can be observed with many other bacterial types. The transforming principle in all these experiments is DNA. Everything that has happened since then, as we shall see in the lectures devoted to genetics, has confirmed the conclusion cautiously put forward by Avery that genes consisted of DNA, or at least contained DNA, and that the DNA of each gene was specific.

The DNA that is active as transforming principle has the usual double-stranded, double helix structure. Since double-stranded nucleic acid molecules are very fragile, this DNA in cell extracts is usually in relatively small fragments, 10^6 to 10^7 in molecular weight. Once a fragment of DNA enters a cell, it can associate with the corresponding DNA sequence of the cell, in a way that we shall discuss later. Each gene coming from the outside can then match the corresponding gene in the cell and has a chance to replace it.

For the time being, what we want to keep in

mind is the molecular basis of all these interactions, from the pairing of DNA strands with
complementary DNA or RNA strands to the replace-
ments of a gene by a transforming DNA. All these
events involving nucleic acids, as well as all
those involving protein structure, function, and
mutual interaction are determined by the sequence
of elements--nucleotides or amino acids. Once
the sequence is formed, these molecules assume
three-dimensional configuration determined by the
attractions and repulsions of the side groups,
each chain being arranged so as to maximize
attractions and minimize repulsive interactions.
In nucleic acids, the double helix represents the
most stable structure because it maximizes the
number of bonds between chains. Thus all DNA and
all complementary sets of nucleic acid chains
have a common optimal structure, the double helix.
(Proteins, instead, are left each to find its own
final structure by the nature of its amino acid
sequence--a sequence, remember, that is not
accidental but is the outcome of the unceasing
trial-and-error testing by natural selection.)

Messenger RNA

The cellular RNA is made as a transcript of
the DNA. When RNA is extracted very gently from
broken cells, its molecules fall into several
classes.

One class of RNA is the messenger RNA. DNA
does not directly serve as template for protein
synthesis. Instead, it sends messenger molecules
of RNA, transcribed on one strand of DNA, to
direct the synthesis of proteins. The sequence
of bases in messenger RNA (mRNA) codes for the
sequence of amino acids in proteins. As we shall
see later, one molecule of mRNA may be the transcript of one or more genes: that is, it may have
the specification for one or more polypeptide
chains.

Transfer RNA

Another class of RNA is transfer or tRNA, the
set of adaptor molecules that combine with
specific amino acids (amino acyl tRNA) and line
them up on the messenger so that they can be
added one-by-one to the growing protein chains.

Ribosomes

The amino acid never actually "sees" the gene or the messenger; the recognition is made through the tRNA.

To make proteins, the mRNA and the appropriate amino acyl tRNA's cannot merely be flopping around in space. They come together in a well-defined fashion on structural elements called <u>ribosomes</u>, each of which contains about one hundred different proteins. Some of these function as enzymes in assembling the protein chains. Each ribosome

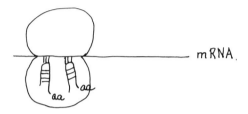

ribosome

also has 2 molecules of ribosomal or rRNA. Nobody knows yet how rRNA participates in "protein synthesis"; but a single break in one of the rRNA molecules renders that ribosome useless. Each ribosome looks like a little mushroom in the electron microscope; in its groove, mRNA and tRNA come together according to the genetic code. The mechanisms of all these processes we shall see later in these lectures.

Before leaving nucleic acids, one question: Why are they called acids? Because they have phosphate groups, which in solution release hydrogen ions.

5'nucleotide,

two acid group in phosphate

phosphate in nucleic acid,

one acid group

Each nucleotide in RNA or DNA has a strong acid group ($pK_a \approx 1$). These groups generally exist not as acids but as salts of such positive ions as Mg^{++}, K^+, or Na^+. In cells, the salts are generally with Mg^{++} or with organic cations, which form stronger bonds with the nucleic acids. But an important consequence of the charge of nucleic acids is that they have affinity for positively charged proteins, some of which are associated with DNA in chromosomes.

Further reading

Watson's Chapter 6, "The Concept of Template Surfaces," will bring additional insight on the relation between molecular structure and function.

Part II

BIOCHEMISTRY

Lecture 5

Metabolism

Now that you have learned what proteins and nucleic acids are made of, you may ask how they are built. What are the precursors that provide the pieces? Where did the energy to assemble them come from? How are the pieces put together?

This involves us in the dynamics of cellular processes as contrasted with the statics of the component materials. This is metabolism: the complex of chemical reactions that convert foods into cellular components, provide the energy for synthesis, and get rid of used up materials. Metabolism is artificially thought to consist of anabolism, or building up, and catabolism, or breaking down activities. But, as you shall see, building and tearing down are inextricably inter-woven in cellular affairs just as they are in human affairs.

All large molecules are assembled from small organic precursor molecules. The ultimate source of these is food. Some organisms, including E. coli and plants, can make all their organic compounds from simple inorganic substances plus a few organic compounds which they must obtain from the environment (or make by photosynthesis). Mammals and most other animals have to be given all sorts of organic compounds, such as the amino acids and the precursor molecules of nucleic acids, since they cannot themselves make them. The external supply, however, can only consist of molecules that can enter the cells. Before they can be built into proteins or nucleic acids they may be drastically altered, and this must be done by chemical reactions inside the cells.

The other requirement for the synthesis of macromolecules, besides the precursor molecules, is energy. Energy is used up or made available in the reshuffling of chemical bonds. The build-ing of large molecules from small ones usually consumes energy, which must be furnished in some way by other chemical reactions that release energy. For example, consider the conversion of glucose to lactic acid, a reaction that provides energy to muscles. A molecule of glucose is equivalent on paper to two molecules of lactic acid:

$C_6 H_{12} O_6$ $2 \times C_3 H_6 O_3$

glucose   lactic acid

However, the combustion or burning of a certain
amount of glucose yields more energy than the
combustion of the equivalent amount of lactic
acid. The difference is about 30 kcal/mole (that
is, 30,000 more calories for every mole of glucose burned than for every two moles of lactic
acid) and is the difference between the sum of
the energies released in separating all atoms in
a molecule of glucose and converting them to CO_2
and H_2O and of the energies released in doing the
same for two molecules of lactic acid--in other
words, the difference in bond energies. Thus
glucose has more bond energy than 2 lactic acids,
and the conversion of one mole of glucose to two
moles of lactic acid can make some energy avail-
able--provided a cell has the machinery to convert
it into a usable form.

Enzymes

    Is anything else needed to make large molecules?
Suppose a muscle needs to split sugar to lactic
acid, as it does when it contracts anaerobically,
that is, with little or no oxygen. Even though
the series of reactions that converts glucose to
lactic acid has a tendency to proceed spontaneously, it does so almost infinitely slowly. To speed
it up the cell has catalysts called enzymes. Each
enzyme is a protein molecule that specifically
catalyzes a particular reaction. We have already
seen this for the lysozyme-catalyzed split of the
cell wall polysaccharide of bacteria.

    Consider now the following reaction:

$$ATP \xrightleftharpoons[-H_2O]{+H_2O} ADP + H_3PO_4$$

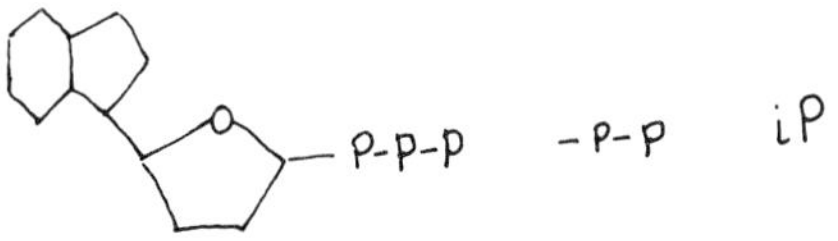

ATP (adenosine triphosphate) is a very important
small molecule: it is the energy currency of the
cell. Placed in water ATP splits to form ADP
(adenosine diphosphate) plus phosphoric acid (iP
for inorganic phosphate). ADP has only two phosphates and is formed by release of the last phosphate from ATP. This reaction, the hydrolysis of
ATP, has an enormous tendency to proceed from

left to right: at equilibrium the ratio ADP/ATP
is about 10^5:1. Yet a solution of ATP will remain
almost unchanged for months in the refrigerator
or for days at room temperature. Despite the
strong tendency to proceed, the reaction spontaneously proceeds very slowly. This can be described
in an energy diagram:

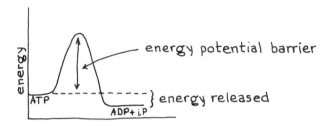

There is a great deal of energy released in
going from ATP to ADP, but there is little chance
for an ATP molecule to be converted to an ADP in
any one interval of time because there is an
energy potential barrier. What does this mean?
In a population of molecules of ATP (or of any
other substance) there is a distribution of
energies, because individual molecules are subject to collisions with other molecules, internal
vibration of atoms along their bonds, etc.

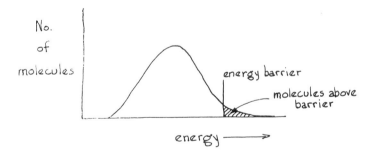

Most molecules have energy near the average energy,
not enough to jump over the potential barriers.
Only a small percentage has enough energy to make
the transition over the barrier. To put it in
chemical rather than physical terms, to undergo
the reaction a molecule (for example, ATP) must
become "activated," that is, distorted to a transition state from which it then glides into the new
structure (for example, ADP + H_3PO_4). The higher
the temperature, the higher is the thermal motion,

resulting in more collisions: therefore, more
molecules have energies above the potential
barrier and the reaction goes faster (see pages
387B-400B).

What a catalyst such as an enzyme does is to
form with the substrate molecules a complex that
distorts the molecules forcing them into a state
close to the activated transition state at the top
of the energy barrier. The result is like that
achieved by raising the temperature: more molecules cross the potential barrier and the reaction goes faster. An enzyme does not change the
difference in energy between the substrate and
the product; it reduces the effective potential
barrier, that is, the excess energy needed to
make the transition.

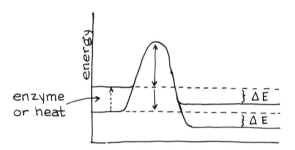

In our diagram, there are two critical parameters; (1) the energy difference between substrate and product, which determines in which
direction the reaction will proceed; and (2) the
potential barrier, which controls the rate of the
reaction. The two parameters are not related to
one another.

An enzyme acts on specific bonds in the substrate molecules: for example, in the hydrolysis
of ATP the enzyme ATPase operates primarily on
the P-O-P bond of ATP and on the HO-H bond of the
water molecule. Of course, the surface groups of
the enzyme must recognize the whole ATP molecule,
otherwise the enzyme could not be specific for
that reaction. (It must also recognize the product substances, ADP and H_3PO_4, otherwise the
reaction could not be reversibly catalyzed.

Enzyme kinetics

For any chemical reaction S ⇌ P (S = substrates;
P = products), the rate of decrease in the amount
of substrate is proportional to the amount present:

$-d[S]/dt = d[P]/dt = k[S]$.

The rate constant k depends on the temperature,
on the presence or absence of the appropriate
enzyme, on the efficiency of the enzyme, and on
its concentration. Some enzymes are very efficient, others are much less so. A mutation that
produces a change of one amino acid in an enzyme
may alter or completely destroy the enzyme's
efficiency, for example, if it changes the configuration of the active site.

   How do we interpret the action of an enzyme?
According to the standard theory, formulated by
Michaelis and Menten, the enzyme E first combines
with the substrate to form an enzyme-substrate
complex ES; this complex then separates, releasing
the product P plus the intact enzyme. The enzyme
is not used up in the reaction.

$$S + E \underset{k_2}{\overset{k_1}{\rightleftharpoons}} (ES) \underset{k_4}{\overset{k_3}{\rightleftharpoons}} E + P$$

The substrate (and the product) can be one or more
substances. Each of the reactions above has a
certain rate, k_1, k_2, k_3, k_4. This complicated situation is simplified by assuming that the reaction
P + E → PS does not occur at a significant rate.
(This is reasonable at the beginning of the reaction when there is a lot of substrate and very
little product. Also, in biological systems the
product is often immediately used for something
else and therefore not available for the reverse
reaction. For example, an amino acid once synthesized quickly becomes incorporated into proteins.)

   Using the above one can derive for the velocity
V of the reaction the equation:

$$V = \frac{K_3[E]}{1 + K_m/[S]} = \frac{V_{max}}{1 + K_m/[S]} .$$

From this equation it is apparent that the velocity
of an enzyme-catalyzed reaction is a function of
the concentration of the enzyme [E] and the concentration of the substrate [S]. The following is a
plot of the velocity for a given concentration of
enzyme (the constants V_{max} and K_m are characteristic for each enzyme):

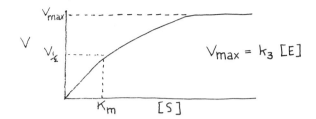

The curve levels off at a maximum velocity V_{max}.
This is when every molecule of enzyme is working
as fast as it can because there is plenty of substrate. From this curve, if we know how much
enzyme we have in our reaction, we can figure out
the turnover number of that enzyme, that is, the
maximum number of substrate molecules that one
enzyme molecule can handle per unit time at that
temperature. For different enzymes, turnover
numbers range from several thousands to about one
per second. [Note: A bacterium such as E. coli
contains 3,000 or 4,000 enzymes. Each enzyme is
coded for by one or a few genes: thus the program must contain at least 3,000 to 4,000 genes.]

Is the amount of a particular enzyme in the
cell cytoplasm fixed? Not at all. The level
changes depending on the cell's need for it. In
bacteria, some enzyme levels are precisely regulated depending on the quantity and quality of
the foods available. In mammalian cells the regulation of certain enzyme levels depends on hormones, but we know much less about it than we know
about enzyme regulation in bacteria. More of
this later.

Inhibition

In cells, not only the amounts, but also the
activity of enzymes is regulated. The regulator
substances fall into two categories: activators
and inhibitors. Inhibitors decrease enzyme activity, either by competing with the substrate for
the active site of the enzyme and in that way decreasing the effective turnover number (competitive inhibitors), or by changing the configuration
of the enzyme and making it less active. An instance of this second type is feedback inhibition.
For example, an amino acid is made by the following pathway in which each arrow indicates an
enzyme reaction.

$$A \xrightarrow{1} B \xrightarrow{2} C \xrightarrow{3} D \xrightarrow{4} X$$
regulation

The level of X in the cell controls the activity
of enzyme 1 of the pathway. The more X is present,
the less active the enzyme is. If lots of X is
added to the medium external to the cell, the
enzyme becomes very inactive. When external X decreases, the enzyme becomes active again. Even in
the test tube X inhibits just that enzyme. Likewise, there are <u>activators</u>, which increase the
activity of enzymes when they combine with them.

This kind of regulation, very frequent at least
in bacteria, is called <u>allosteric</u> because the
regulator combines with the enzyme at a site other
than the active site. This is demonstrated by the
effects of the allosteric regulators on the
kinetics of enzyme action. Also, in some cases of
enzymes consisting of several separable protein
subunits, it is actually found that the active
site is in one subunit and the attachment site
for the regulator in a different subunit!

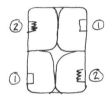

1= active site

2= regulatory site

There is hardly any need to stress how efficient
such regulation can be, sparing to a cell the making of a substance in excess of what it needs or
speeding up the synthesis of one it needs. One
should reflect, rather, on the almost uncanny
subtlety of the natural selection process that
has imprinted in the genes, not only the code of
making usable enzymes but enzymes whose structure
has specific sensors for regulatory information.
These sensors monitor the level of substances that
are the ultimate, distant product of the whole
series of enzyme reactions of which the regulated
enzyme is only the first one.

Coenzymes

In catalyzing a reaction, the enzyme molecules
are not altered and can be used more than once.
One enzyme molecule can catalyze the chemical conversion of a very large number of substrate molecules. Therefore we expect, correctly, to find
that enzymes are present in relatively small molar
concentrations (on the order of 100 to 10^6 mole-

cules of a given enzyme per cell), while substrates
are found in much larger quantities. We say that
enzymes are present in catalytic amounts. There
is another class of substances that are present in
catalytic amounts, although they are not themselves
catalysts. These are the coenzymes, small molecules that take part in intermediate reactions of
several chemical pathways. For example, a substance can be oxidized to another substance by
donating a pair of electrons to a coenzyme. Then
the coenzyme is regenerated by transferring the
electrons to something else:

$$\text{substrate} + \text{coenzyme} \longrightarrow (\text{substrate} - 2e^-) + (\text{coenzyme} + 2e^-)$$

$$NAD^+ \xrightarrow{\ +2e^-,\ +H^+\ } NADH$$

$$NADH + \text{acceptor} \longrightarrow NAD^+ + (\text{acceptor} + 2e^-)$$

$$NADH + CH_3\,CO\,COOH \longrightarrow NAD^+ + CH_3\,CHOH\,COOH$$
$$\qquad \text{pyruvic acid} \qquad\qquad\qquad\qquad \text{lactic acid}$$

The transfer of electrons from NADH to pyruvic
acid is the source of lactic acid, as we shall
soon see in detail. The total amount of the coenzyme needed in the cell is small because it acts
only as a carrier of electrons and is not used up,
but is regenerated by some other reaction.
    Practically all coenzymes are related to vita-
mins. Vitamins are substances that certain animals,
including man, need to receive in small amounts in
their food because they cannot manufacture them in
their own body. The molecule of NAD^+, for example
(see page 402B), contains a molecule of the vitamin nicotinamide. (Nicotinamide is now called
niacin because the name nicotinamide suggested a
relation to smoking and this bothered the manu-
facturers of breakfast cereals.) Why then do we
eat nicotinamide rather than NAD^+? Simply because the coenzyme NAD^+ is an electrically charged
molecule and therefore cannot enter cells. Nicotinamide is uncharged and can enter and, once inside,
it is converted to NAD^+. Other vitamins, such as
riboflavin, thiamine, and pantothenic acid, are
incorporated into coenzymes that participate in
the reactions of cellular metabolism.

Further reading   Lehninger's Chapter 8, "Enzymes: Kinetics and
                       Inhibition," will complement this lecture.

Lecture 6

Chemical
energetics

We have previously drawn an energy diagram for a
reversible chemical reaction:

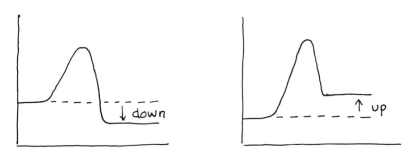

All reversible reactions tend to proceed in the
direction that decreases the energy of the parti-
cipants. As you already know, even in this direction (let us leave the term energy undefined for
a little while) the reaction may not proceed very
quickly because very few molecules overcome the
potential energy barrier and make the transition
toward the lower energy state.

Suppose now you want to go in the opposite
direction; that is, you want to convert a substance or substances at one energy level into
other substances with higher evergy level. For
example, you want to convert a mixture of amino
acids into a protein. Such a reaction requires
energy. It is called an endothermic reaction and
will not go spontaneously, even with enzymes to
catalyze it. Energy must be supplied. In general,
all series of reactions involved in the synthesis
of large molecules are endothermic. For example,
the conversion of sugars to sugar-phosphates used
to make polysaccharide requires an input of energy.
The simple reaction cannot proceed:

$$glucose + iP \not\longrightarrow glucose\text{-}6\text{-}P$$

cannot go

How is energy made available in cells? In what
form is it stored? In what form and by what reactions is it used?

Chemical potential

For a chemical reaction, the chemical potential
(or potential difference) that drives it involves
two things: first, the intrinsic properties of
the substrate and product molecules; second, their

concentrations. If a reaction $A + B \rightleftharpoons C + D$ is
reversible, the direction in which it goes must
depend on the properties of A,B,C,D and on how
much of each there is. The more A and B one adds,
the more C and D are formed, and vice versa.
Also, if C or D is removed as fast as it is made
the reaction is driven to the right. We shall
deal with this quantitatively in a moment.

When a cell needs to make a reaction go in an
energetically unfavorable (endergonic) direction,
it uses the mechanism called coupling: it converts the substrate to something with higher
chemical potential by using another reaction that
is energetically favorable, that is, exergonic.
The chemical potential of the modified substrate
is higher than that of the product so that the
reaction will now tend to proceed spontaneously in
the desired direction. For example, the conversion glucose + iP $\rightarrow$ glucose-6-P + H_2O requires
energy. The reaction $ATP + H_2O \rightarrow ADP + H_3PO_4$
releases more energy. They are coupled (on an
enzyme surface) into the overall reaction: glucose
+ ATP $\rightarrow$ glucose-6-P + ADP, which is still exergonic.
Here, to carry out the reaction the cell uses a
phosphate group that is trapped into the ATP molecule, which acts a donor of phosphate to glucose.

Cells use similar mechanisms to make all sorts
of energetically unfavorable reactions proceed.
In protein synthesis, for example, the first step,
aa + ATP $\rightarrow$ AMP-aa + iPP (inorganic pyrophosphate),
is exergonic and converts the amino acid (aa)
into a new form, which still retains some of the
high chemical potential and serves as a donor of
amino acid for making protein. Note that ATP was
used in both the above examples of coupled reactions. In the first case it donated one of its
phosphate groups to glucose and thereby increased
the reactivity of the glucose so that it could be
used for other reactions. In the second case, ATP
donated another part of its molecule to an amino
acid.

ATP

ATP is the most important of a group of compounds that have traditionally been called high-
energy compounds. More correctly, it is the prototype of a group of molecules with high group donor
potential, that is, molecules that release much
energy when they donate some parts of themselves
to water or to other acceptor molecules.

ATP

AMP

ADP

~ "high energy bonds"

High-energy bonds

(Remember that all these reactions proceed very slowly without the appropriate enzymes!) The parts donated may be atoms or groups of atoms or, in some cases, just pairs of electrons.

In ATP, there are at least two binding places that have a tendency to participate in the donation of parts of the molecule: AMP $\perp$ P $\perp$ P. Traditionally, but incorrectly, these are referred as <u>high-energy bonds</u> and represented by the symbol $\sim$: AMP $\sim$ P $\sim$ P. The energy (that is, the chemical potential), the tendency to transfer chemical groups, is in the molecule as a whole, not in this or that bond. The same is true, for example, of NADH: it has a high electron donor potential and can therefore "reduce," that is, transfer electrons to a variety of substances.

Since the chemical machinery of the cell is driven by substances with high group donor potential, the next question is, How can these substances be made? Cells use only two kinds of energy: (1) light energy, trapped and used by plants and some bacteria for photosynthesis; and (2) chemical energy, that is, the energy held in the bonds of various chemicals. Cells do not use thermal or electrical energy because they don't have thermal or electrical converters. Thermal potential (that is, temperature) affects the rate of chemical reactions, but does not provide any energy. [What about the electrical signals of nervous impulses? They use energy in the form of ATP to generate electric potentials in the membrane of nerve cells and fibers.]

Free energy

How can we measure the energetic budget of a cell? The change ΔG in usable energy (that is, chemical potential) in a reaction is described by the equation

$$\Delta E = T\Delta S + \Delta G \ ,$$

where ΔE = overall change in energy, T = absolute temperature, ΔS = change in entropy (which measures changes in the order of the system), and ΔG = change in free energy or chemical potential. The chemical potential (or Gibbs free energy) is the amount of energy that is available to do work. The change in free energy for a reversible reaction, A + B $\rightleftharpoons$ C + D is given by

$$\Delta G = \Delta G_0 + RT \ln \frac{[C][D]}{[A][B]} \ ,$$

where R = gas constant, T = absolute temperature, and ΔG_0 is a constant characteristic of the substances involved.

Suppose we let the reaction go to equilibrium. Then the ratio [C][D]/[A][B] equals K_{eq}. By definition at equilibrium there are no changes in concentration and $\Delta G = 0$. Therefore

$$\Delta G = \Delta G_0 + RT \ln K_{eq} \ ,$$

$$\Delta G_0 = -RT \ln K_{eq} \ .$$

This relationship makes it possible to measure the change in free energy for a given chemical reaction if one can measure the equilibrium concentration of products and substrates.
Rule: <u>A chemical reaction always goes in the direction that tends to decrease the free energy</u> (the direction for which $\Delta G < 0$).

A reaction will not go in the opposite direction unless it is pushed or pulled. How can one push or pull? By varying the concentrations of reactants and/or products. One way, as I mentioned earlier, is to remove one product. For example, if the reaction is A + B $\rightleftharpoons$ C + D another reaction may remove D as soon as it is made. Alternatively, another reaction can produce more and more A or B. Increased production of C or D would push the reaction backward.

It is very important to remember that the two factors that determine the change in free energy in a reaction are the intrinsic properties of the substances and the concentrations of the reactants and products, because it helps us to understand why certain reactions go. For example, in the conversion of glucose to lactic acid, which serves to produce ATP for muscle contraction and many other processes, there is a series of reactions, some

of which are uphill (energy-requiring, ΔG > 0) and
some which are downhill (energy-releasing, ΔG < 0).
The uphill reactions can proceed only because the
products are continuously removed for use in later
downhill reactions. If you have to go over a
series of hills

the only thing that matters is that the initial
level is higher than the ultimate level so that
the process as a whole represents a descent.
[Water in a completely filled tube can go up and
downhill provided the initial pressure is higher
than the final pressure.] The ΔG for the entire
series of reactions is the algebraic sum of the
ΔG's of the individual reactions. ΔG is the variable on the ordinate of the energy diagrams used
on pages 51, 52, and 57).

The relation of the concentration of a substance to the free energy becomes clearer when we
consider active transport across a membrane, that
is, the transfer of a substance from a region of
lower concentration to one of higher concentration.
Energy is needed to overcome the gradient of concentrations (since the substance would tend to
flow down the gradient till the concentrations are
equal):

$$\Delta G = RT \ln \frac{conc_{out}}{conc_{in}} \, .$$

If $conc_{in}$ is higher than $conc_{out}$ the ΔG for transport inward is positive and energy is required.
In living cells this energy comes from ATP or
other substances with high group donor potential,
but how it is actually used for transport remains
obscure. [One would surmise that energy is used
to distort or rotate molecules of specific trans-
porter proteins.]

Substances like ATP, with high group donor
potentials, have ΔG_0 for hydrolysis lower than
-7 kcal/mole. The transfer of one phosphate or of
two phosphates (P-P) or of the AMP group to water

or to some other substance releases more than 7
kcal/mole. For example, in the hydrolysis of ATP
in water,

ATP + H$_2$O → ADP + H$_3$PO$_4$,

8,000 calories are released as heat (measurable
in a calorimeter) for every mole of ATP that disappears. Sometimes, ATP donates a phosphate to
another substance to generate a compound that itself has a high group donor potential. For example, <u>acetyl phosphate</u>, formed in the reaction

ATP + acetic acid ⇌ ADP + acetyl phosphate

is itself capable of donating phosphate to other
substances. In general, any phosphate group that
is attached to an acid through an <u>anhydride</u> bond
-(CO)-O-P is very reactive. [An anhydride bond
is formed by the joining of two acids face to
face with removal of a molecule of water. It has
a high tendency to be split by water to reform
the two acids or to donate one of the acid groups.]
   When ATP donates a phosphate to an alcohol
(-C-OH) as in the already familiar reaction

glucose + ATP → glucose-6-phosphate + ADP

the substance that is formed does <u>not</u> have a high
group donor potential. For example, if glucose-6-P
were to hydrolyze in water it would release only
about 3 kcal/mole; in fact, it has little tendency
to do so.

Activation
   Finally, consider once more the situation in
the <u>activation</u> of amino acids for protein synthesis. First the amino acid reacts with ATP to
form a compound with high group donor potential:

$$R-\underset{\underset{NH_2}{|}}{\overset{\overset{H}{|}}{C}} - COOH + ATP \longrightarrow AMP-O-\overset{\overset{O}{\|}}{C}-\underset{\underset{NH_2}{|}}{\overset{\overset{H}{|}}{C}}-R + iPP$$

amino acyl-AMP

Right on the <u>activating enzyme</u> where it was
formed, the amino acyl-AMP reacts directly with
the specific adapter molecule tRNA:

amino acyl ∿ AMP + tRNA → amino acyl ∿ tRNA + AMP.

The aa∿tRNA then serves as donor of the amino acid

for protein synthesis. Notice the series of
tricks: The amino acid itself cannot be added
directly to the growing protein because there is
not enough energy. In the steps ATP + aa → aa ∿
AMP + iPP the needed energy is contributed by the
split of ATP. But the aa∿AMP has no way of
recognizing <u>where</u> the molecule should go in the
protein (because all transfer of information must
be between nucleic acids). So the amino acid is
transferred to a specific tRNA with the preserva-
tion of the high-energy bond. The amino acyl∿tRNA
molecule is not only energetic but also smart: it
can recognize the code on the mRNA for inserting
the amino acid into a protein. Thus in this process we have added both energy and intelligence.
The remarkably smart fellow in this whole process
is the activating enzyme that catalyzes the reactions from amino acid to aa∿tRNA: it must recognize specifically a given amino acid, its appropriate tRNA, as well as ATP!

Pyrophosphate split   Another important feature remains to be considered. The reactions

ATP + amino acid → amino acid ∿ AMP + iP∿P and

amino acid ∿ AMP + tRNA → amino acid∿tRNA + AMP

do not really release much energy. There are
equal numbers of "high-energy bonds" on both sides.
Left to itself the overall reaction would actually
tend to go backwards. What prevents this is that
in all living cells there is an enzyme that carries
out the reaction called <u>pyrophosphate split</u>.

iP∿P + H_2O → 2iP

The split destroys a high-energy bond and continuously removes one of the products of the first
reaction and prevents the reverse reaction from
occurring. To put it more quantitively, the coup-
ling of the activation reaction ATP + amino
acid ⇌ AMPaa + iPP (ΔG_0 = +0.5 kcal/mole) with
the pyrophosphate split iPP → 2iP (ΔG_0 - 7.5
kcal/mole) renders the overall reaction strongly
exergonic. The pyrophosphate split is one of the
neat tricks of living cells economy.

Further reading   Pages 387B-399B clarify further the thermo-
dynamic aspects. The eager reader will find
Klotz's little book exciting. See also Watson,
Chapter 5, "Coupled Reactions and Group Transfers."

Lecture 7

Oxidation reactions

What are the actual chemical mechanisms by which cells get energy from food? They consist of series of reactions that include one or more oxidation reactions, which are coupled with reactions that generate high-energy compounds, especially ATP. An oxidation reaction is one in which electrons are transferred from one compound to another. A reduction reaction is one in which electrons are added to a compound. Naturally, every oxidation reaction is also a reduction reaction. For example, an aldehyde can be oxidized to the corresponding acid.

$$CH_3 - CHO + H_2O \rightarrow CH_3 - COOH + 2H^+ + 2e^-$$

The acid is more oxidized than the aldehyde. NAD^+ is reduced to NADH which has two more electrons.

$$NAD^+ + 2e^- + H^+ \rightarrow NADH$$

The two above reactions are combined in an oxidation-reduction reaction. The aldehyde donates two electrons to NAD^+, with the result that the aldehyde is oxidized to acid and the NAD^+ is reduced to NADH. Generally, in oxidation-reduction reactions one pair of electrons is transferred from one substance to another. An aldehyde can be considered as having a high electron donor potential (see page 400B).

These reactions often result in the release of amounts of energy large enough (that is, they have a negative enough G_0) that they can serve to generate compounds like ATP. In the oxidation of an aldehyde, for example, the acid may, before leaving the enzyme, pick up a phosphate from the medium to yield a high-energy compound:

$$CH_3 - CHO + H_2O + NAD^+ \rightarrow CH_3 - COOH + NADH + H^+$$

This in turn can then transfer the phosphate to ADP to produce a molecule of ATP.

$$R - CHO + NAD^+ + iP \rightarrow R - COO \sim P + NADH.$$

Note that, if the acid intermediate is released into the medium before the phosphate is attached to it, no high energy bond is formed and the energy released by oxidation is lost. The enzyme preserves the activated state of the substrate so that it can serve as an acceptor of the phosphate.

Not all oxidation reactions permit the production of high-energy compounds. For example, yeast produces ethanol by reducing acetaldehyde and oxidizing NADH:

$$CH_3 - CHO + NADH + H^+ \rightarrow CH_3CH_2OH + NAD^+$$

Here the enzyme is unable to use the reaction for
high-energy-bond production.

To summarize: oxidation-reduction reactions
are those in which pairs of electrons are transferred. These transfers of electrons, depending
on the nature of the substrate, may or may not
represent changes of energy sufficient to generate
high-energy products. If they do not, then whatever energy is released is dissipated in heat.
When energy release is sufficient, the oxidized
substrate may become the acceptor for other groups
and generate compounds of high group donor potential. Thus the energy of chemical bonds is made
available, by oxidation reactions, to the chemical
machinery of cells.

lycolysis

One of the most important pathways used by the
cell to produce energy is glycolysis. It occurs
in all plant and animal cells and in most bacteria.
It is the device that was used very early (at
least a couple of billion years ago, long before
photosynthesis was invented) to produce ATP from
organic substances. Glycolysis, the splitting of
sugars to release energy, includes some of the
reactions we have discussed.

A representation of the pathway of glycolysis
is shown on the facing page. More schematically,
this can be represented as follows:

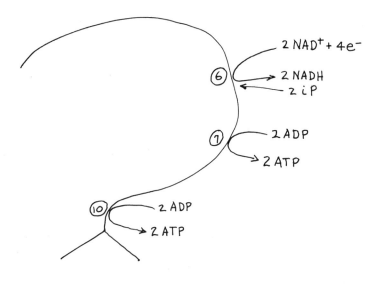

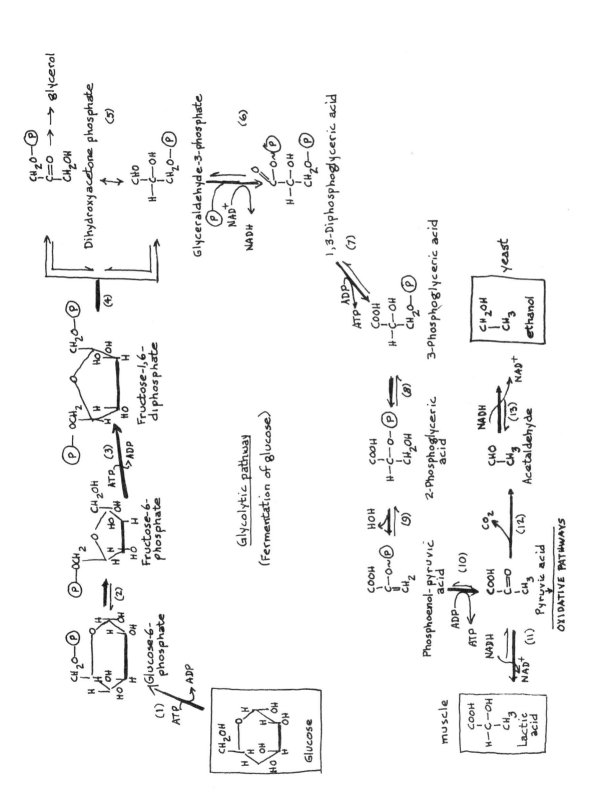

Glycolytic pathway

(Fermentation of glucose)

The circled numbers are those of the reactions in the complete scheme.

Glycolysis that terminates in converting sugar into some organic products is called fermentation. Some cells, such as yeast, produce ethanol and CO_2 from the fermentation of glucose; others, such as muscles, ferment sugar to lactic acid.

The first important point to note is that cells do not take one substrate and use it directly for whatever purpose they have. Instead, they use pathways, which are series of reactions, each of which has its own enzyme and uses the product of the preceding reaction. Some reactions in a pathway have positive ΔG_0, but the whole series of reactions must have a net negative ΔG_0. In such a pathway, there are only a few "useful" reactions. In glycolysis only two reactions (reactions 7 and 10) produce ATP.

The second point is the source of energy used to make ATP. This energy comes from the splitting of bonds that held together the glucose molecule. If you burn a mole of glucose you get 676 kcal of energy. Since combustion of two moles of lactic acid yields about 646 kcal of energy, there are 30 kcal available to generate ATP in muscle fermentation.

The third point is how the energy is made available. There is in glycolysis a key reaction, reaction 6, in which an aldehyde is converted to an acid. This is the oxidation reaction we already know and releases lots of energy as two electrons are passed from the aldehyde to the coenzyme NAD^+ together with one H^+ from water. Before leaving the enzymes, the acid that is formed picks up a phosphate from the medium to give an anhydride, 1,3-diphosphoglyceric acid, with so much phosphate donor potential that in two steps, reactions 7 and 10, it transfers two phosphates to two ADP to make two ATP.

In the early reactions of glycolysis a six-carbon sugar is split to form two three-carbon compounds, which are interconvertible (reaction 5). Thus the reactions in the latter part of the pathway (from reaction 6 on) occur twice for each molecule of glucose, and so for each molecule of glucose there is a production of 2ATP + 2ATP = 4ATP. However, looking at the pathway, you see that this is not all net gain. Before undergoing

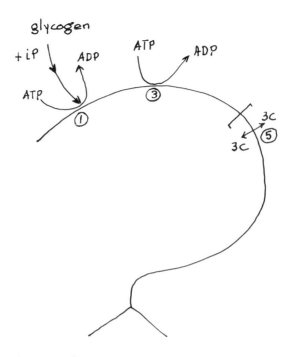

glycolysis glucose must be phosphorylated twice
(reactions 1 and 3), so that 2 ATP's are used up
(that is, converted to ADP) in order to get the
reactions going. Therefore, the net energy gain
is 4ATP - 2ATP = 2ATP per molecule of glucose converted to 2 molecules of lactic acid (or ethanol
and CO_2 in alcohol fermentation by yeast).

Note that glycolysis as discussed above occurs
without intervention of oxygen. All cells that do
not have direct access to oxygen--for example,
anaerobic bacteria or muscle cells working under
lasting effort--use this fermentation process or
some comparable process for energy production. A
fermentation is a series of reactions by which an
organic compound gets converted to other organic
compounds with the production of ATP. Pasteur
defined fermentation by yeast as "life without
air." A bacterium like E. coli, given some sugar,
can get all the energy it needs by fermenting the
sugar: it can do without oxygen. In muscle,
glucose is stored as glycogen and when needed it
is released as glucose-1-P, which then becomes
glucose-6-P (bypassing reaction 1 of glycoysis).

What would happen if the pathway stopped at
pyruvic acid after reaction 10? Why are the
reactions following pyruvic acid in the fermenta-
tions important?

Regeneration of NAD^+ Let us look at the pathway once more. These reactions are essential because they regenerate NAD^+. In the pathway from glucose to pyruvic acid NADH is formed as a product of reaction 6. Remember that NAD^+ is a coenzyme: it is present in any cell in small quantities; if it were not regenerated it would soon all be converted to NADH and the overall process would stop when half as many moles of glucose had been used as there were molecules of NAD^+. For glycolysis to proceed, therefore, something must accept the electrons back from NADH. This substance is pyruvic acid in muscle (reaction 11) or its derivative acetaldehyde in yeast (reactions 12 and 13). The lactic

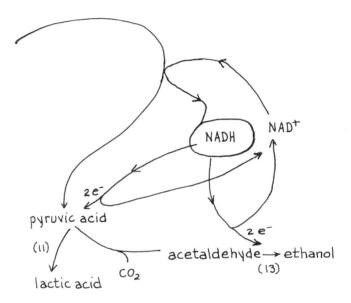

acid, or the CO_2 and ethanol, are excreted unused from the cells. In general, for every 2 molecules of ATP produced from sugar in the absence of oxygen, you must have a molecule that accepts 2 electrons from a molecule of NADH.

Respiration

If, however, oxygen is introduced, the situation is radically changed. The obligation to perform fermentation--that is, to make and use organic molecules such as pyruvic acid or acetaldehyde electrons to regenerate NAD^+--disappears, because oxygen itself is an excellent acceptor of electrons. The electrons of NADH can be transported to oxygen by a series of aerobic reactions that produce a hell of a lot of ATP. In contrast to the anaerobic fermentation, which produces just 2 moles of ATP per mole of glucose, the aerobic

process called <u>respiration</u> can reproduce 36 moles
of ATP per mole of glucose or 18 times as much
energy! Thus, oxygen increases enormously the
efficiency of the energy-producing process, for
two reasons: (1) it is no longer necessary to
use pyruvic acid to oxidize NADH and to make useless products that are excreted; instead, pyruvic
acid can itself be oxidized to produce more energy;
and (2) because the stepwise transfer of electrons
to O_2 to generate H_2O produces large amounts of
ATP.

   Before we look more closely at the situation
created by the presence of oxygen, let us consider
some questions of technique. We have been discussing glycolysis, a biochemical pathway. How do
biochemists learn the details of such pathways?
Several techniques are available:
   1. Chemical identification of possible intermediates produced in cell extracts that contain
(hopefully) some or all the enzymes of the pathway.
   2. Establishment of the order of intermediates
in the pathway by determining the kinetics of
their appearance and disappearance after the addition of the earlier substances in the pathway.

Isotope labeling
   3. Isotope labeling, that is, analysis of
label distribution in the product molecules after
addition of a specifically labeled substrate. For
example, if to a yeast extract capable of glycolysis one adds glucose labeled in a specific position, for example, a ^{14}C atom in position 4, one
can isolate various products or intermediates and
test whether and where they are labeled. All the
labeled carbon atoms in the intermediates must
have come from the C atom in position 4 in the
glucose. Note, for example, that in yeast glycolysis the CO_2 that comes off in the conversion of
pyruvic acid to acetaldehyde is derived from the
C atoms in positions 3 or 4. [Can you see why in
glycolysis, as far as products are concerned, C
atoms 1 or 6, 2 or 5, and 3 or 4 are equivalent?]

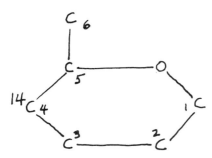

4. The final proof of a pathway is the purification of all the enzymes and the demonstration that they catalyze the respective reactions. These, of course, are only the biochemical techniques. As we shall see later, a potent approach to studying metabolic pathways, both in microbes and eucaryotes, is the genetic study of mutants that are impaired in one or another step of a pathway.

An examination of the ΔG_0 values for the reactions of the glycolysis shown on page 67 reveals that some are negative and others are positive. A plot of these values looks as follows:

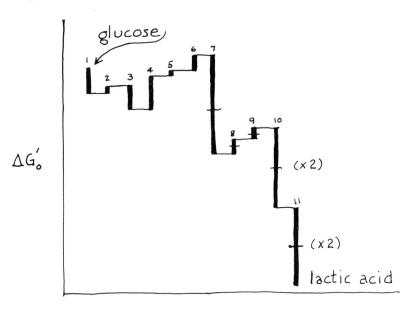

Reaction	K_{eq} (heavy arrow)	$\Delta G_0'$ kcal/mole[*]
1	300	−3.4
2	0.3	+0.5
3	300	−3.4
4	0.00005	+5.7
5	0.05	+1.8
6	0.07	+1.5 (x 2)[†]
7	100,000	−6.8 (x 2)
8	0.2	+1.06 (x 2)
9	0.5	+0.44 (x 2)

Reaction	K_{eq} (heavy arrow)	$\Delta G_0'$ kcal/mole[*]
10	20,000	-5.7 (x 2)
11	30,000	-6.0 (x 2)
12	4,000	-4.7 (x 2)
13	7,000	-5.1 (x 2)

[*]$\Delta G_0'$ is ΔG_0 at pH 7.0, which may differ from ΔG_0 (which refers to pH 0.0) because pH affects ionization.

Balance: glucose + 2 ATP → 2 lactate + 4 ATP ($\Delta G_0'$ - 29.8 kcal/mole).
glucose + 2 ATP → 2 ethanol + 2 CO_2 + 4 ATP ($\Delta G_0'$ - 37.4 kcal/mole).

[†]2 moles for every mole of glucose used.

Since the overall ΔG_0 for a metabolic pathway is the sum of the ΔG_0's for the individual reactions, the total change for the conversion of glucose to lactic acid is about 30 kcal/mole, as we already know. The important point is that, all along, each reaction pulls the preceding one by removing its products. This makes the series of the reactions go, even over the most difficult steps, those that would actually go backwards if the concentrations of the reagents were comparable.

Fermentation of sugars or of other food stuff is clearly not very efficient. It leaves behind substances, such as lactic acid and ethanol, that still have plenty of food value. But it is the best that cells can do when they have no oxygen. Fortunately (for them) most cells contain devices-- mitochondria in animal and plant cells or equivalent structures in the cytoplasmic membrane of bacteria--in which a series of <u>electron carriers</u> accepts electrons from appropriate electron donors and transfers them stepwise from one carrier to the next, storing some of the energy in the form of ATP, until finally the electrons are given up to oxygen.

Further reading

Lehninger's Chapters 15, "Glycolysis," 16, "The Tricarboxylic Acid Cycles," and 17, "Electron Transport and Oxidative Phosphorylation," are hard but excellent and clear supplements to lectures 7 and 8.

Lecture 8

Electron transport

Glycolysis, our friend from the preceding lecture, is a common mechanism by which most organisms extract chemical energy from organic compounds. Under anaerobic conditions, glycolysis or sugar fermentation is by itself sufficient to supply all energy an organism needs. If oxygen is available, however, then a cell can extract much more energy from an organic substrate like glucose for two reasons. First, there are <u>electron transport</u> reactions, by which the electrons released from NADH are transferred to oxygen.

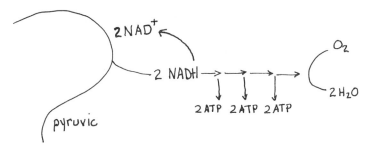

These reactions are coupled with the synthesis of ATP. Second, the use of the electron transport system to accept electrons from NADH regenerates NAD^+; there is no need, therefore, to use up the pyruvic acid to accept electrons, and this pyruvic acid can be used to release more energy as well as useful materials. The presence of O_2 saves the cell from having to waste the products of fermentation.

I already mentioned that the components of the electron transfer reaction chain are present in the membranes of bacteria and of the mitochondria.

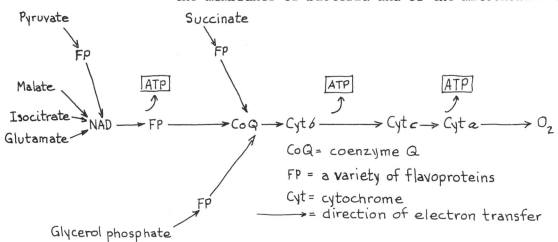

CoQ = coenzyme Q

FP = a variety of flavoproteins

Cyt = cytochrome

⟶ = direction of electron transfer

The electron transport chain includes three or
more types of <u>cytochrome</u>. Cytochromes are proteins, each of which has an atom of iron held
within a heme bound to the protein. In cytochromes, the iron can pick up and lose electrons
readily, passing from the ferrous to the ferric
form and back: $Fe^{+++} + e^- \rightleftharpoons Fe^{++}$. (This is different from the heme of hemoglobin, which only
picks up and releases O_2.)

   In the respiratory chain there are also <u>flavo-
proteins</u>, which contain electron-transfer groups
related to the vitamin riboflavin. They accept
two electrons from NADH or from other donors such
as succinic acid and then pass the electrons one-
by-one to the iron atoms in cytochromes. The proteins of the cytochromes are not all alike, and
each gives to its iron atom a characteristic
affinity for electrons. Thus the electrons go
through a series of jumps down a potential scale
(which corresponds to a free energy scale) until
they get to oxygen.

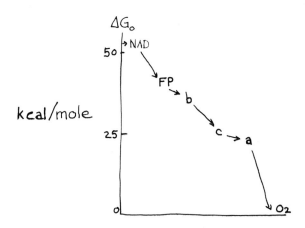

The oxygen picks up electrons from the last cytochrome and H^+ from water and is converted to
water: $O_2 + 4\ e^- + 4\ H^+ \rightarrow 2H_2O$.

   Of course the electrons always move in the
direction for which ΔG_0 is negative. These reactions, therefore, liberate energy; and the essential, although unexplained, feature of electron
transport is that a large part of the released
energy is trapped to form ATP from ADP (<u>oxidative
phosphorylation</u>).

   Thus <u>mitochondria</u> are like little power plants;
they take fuel in the form of electrons from NADH
or other substances and oxidize it in steps. The

Oxidative
phosphorylation

oxidation is coupled to energy production in the same way **as** in a steam power plant where a fuel is used to boil water and the steam is used to run an engine.

Just by using the electron transport chain to regenerate NAD^+, a cell can get 3 molecules of ATP per NAD^+ regenerated, or 6 molecules of ATP per molecule of glucose converted to 2 molecules of pyruvic acid. Thus having oxygen available greatly increases the energy yield from the oxidation of organic compounds.

Life before oxygen

The invention of oxygen as an acceptor of electrons was an enormous step in the evolution of life. This step is believed to have been a relatively late evolutionary development because the primitive atmosphere of the earth was a reducing atmosphere and had little or no oxygen. Oxygen started accumulating in our atmosphere after photosynthesis evolved, that is, at a time when plants such as algae were already present. Probably 2 to 3 billion years ago life changed the chemistry of the earth. Organic matter supposedly had come into being on earth before life got started and piled up as a result of chemical reactions that occurred in a reducing atmosphere. The early organisms could use the available organic matter and get energy by fermenting it by glycolysis or some comparable process. Photosynthesis developed, as we shall see, by adding a new twist to a process related to glycolysis. When oxygen appeared as a by-product of photosynthesis, respiration could develop.

Krebs cycle

Once the cell is free of its slavery to NADH, that is, when it no longer needs to use pyruvic acid to regenerate NAD^+, it can then use pyruvic acid for other purposes. The utilization of pyruvic acid by the cell involves the use of a cyclic series of chemical reactions called the <u>Krebs cycle</u>. [A biochemical cycle is a series of reactions in which a substance A joins with another substance B, then goes through a series of transformations which generate various products and regenerate the original molecule A. Note these features in the Krebs cycle illustration.] The Krebs cycle is a system in which carbon from glucose or other compounds is converted to the fully oxidized product CO_2. This conversion is accomplished by several oxidation reactions that

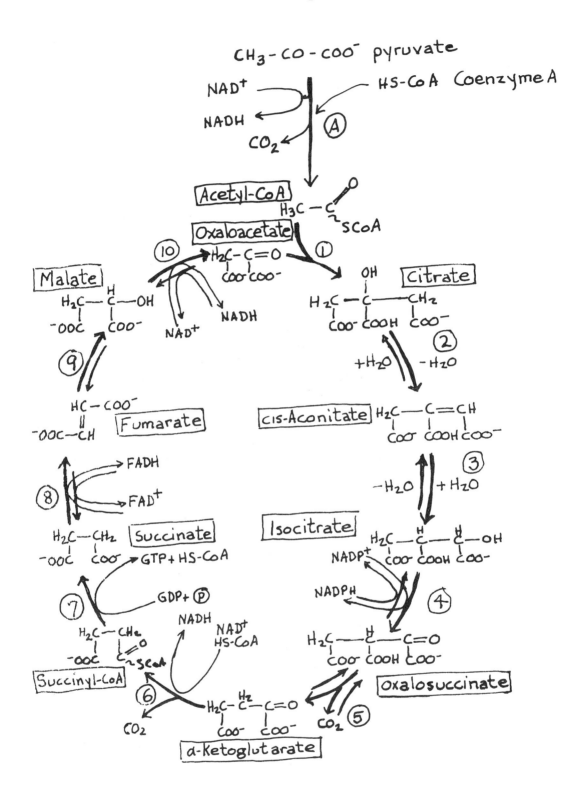

Krebs cycle

produce either NADH or other reduced substances.
This cycle illustrates another important quality
of metabolism: reactions that can yield energy
(catabolism) are generally not handled separately
from reactions that can produce materials for
synthesis (anabolism). The Krebs cycle, for example, serves to produce NADH (which in turn can
give electrons to make ATP), but it also produces
materials from which cells can make amino acids
and other building blocks.

The first step in the cycle is the release of
CO_2 from pyruvic acid; the other two C atoms are
combined with a molecule of the coenzyme HSCoA to
generate acetyl-CoA. This reaction is an oxidation and is coupled with the reduction of NAD^+ to
NADH; the NADH, like that from glycolysis, can
then release its electrons to the electron transport. The formation of acetyl-CoA is a very
shrewd reaction, as we shall see in a moment.
[Note: HSCoA is a coenzyme that contains the
vitamin pantothenic acid.]

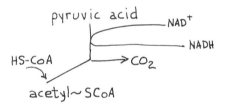

The next reaction in the series involves the
joining of oxaloacetic acid, a four-carbon acid,
with the acetyl group from acetyl-CoA to make a
six-carbon substance.

oxaloacetic acid + acetyl ∿ SCoA → citric acid

+ HSCoA

To synthesize such a compound, energy is needed.
This is where the shrewdness of the previous
reaction becomes apparent. If the cell had simply removed two electrons and CO_2 from pyruvic
acid to form acetic acid, the process would stop
there. Acetic acid and oxaloacetic acids do not
tend to combine: neither of them has high group
donor potential. Instead, the cell stores some
of the energy of the oxidation of pyruvic acid in
NADH and the rest in the high-energy bond in
acetyl-SCoA, CH_3-Co ∿ SCoA, which then makes possible the condensation reaction between oxalo-

acetic acid and acetic acid. Notice the analogy
with reaction 6 of glycolysis: in both cases
some of the energy of an oxidation is salvaged
right on the surface of the enzyme by picking up
something (phosphate, or coenzyme A) to form a
high-energy bond.

Energy yield   In the Krebs cycle there are three reactions
that produce a molecule of NADH (or of NADPH--a
slightly different chemical, but with function
similar to NADH). Each NADH or NADPH can lead to

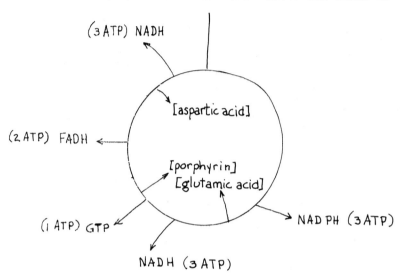

the formation of 3 ATP's by using the electron
transport chain. In addition, one reaction in
the cycle produces a molecule of GTP, which is
equivalent to a molecule of ATP. Another reaction
reduces a flavoprotein FAD^+ to FADH, which also
feeds its electrons into the electron transport,
but a bit too much downstream, so that it can
generate only two molecules of ATP instead of
three from NADH.

Altogether, each turn of the cycle can generate
3 + 3 + 3 + 1 + 2 = 12 ATP molecules in addition
to 3 ATP from the oxidation of pyruvic acid. So
much energy is made because the cycle completely
oxidizes pyruvic acid (or glucose, which is converted to 2 pyruvic acids) to form CO_2 and H_2O;
all of the electrons have been unloaded to oxygen
with the production of energy. The glucose molecule is loaded with chemical energy: its complete
oxidation

$$C_6H_{12}O_6 + 6O_2 \rightarrow 6CO_2 + 6H_2O$$

can release 676 kcal/mole. If one gets 36 moles of ATP from ADP, the ΔG_0 gain is about 350 kcal/mole--a yield better than 50 percent. [The various reactions that oxidize glucose do not take place in spatial relation to each other. The enzymes of glycolysis are "soluble," that is, loose in the cells. Some of the enzymes of the Krebs cycle are soluble, others are in the membranes of mitochondria together with the components of the electron transport. This creates for eucaryotic cells the additional problem of ferrying the substrates in and out of the mitochondria, and some energy is lost in the process. We'll leave that problem to some more advanced course in biochemistry, which I am sure you are all looking forward to.]

Krebs cycle and biosynthesis

Next let us consider a problem which illustrates what I often refer to as the first law of Luria: "you cannot eat your cake and have it." [Incidentally, the second law is, If something is not worth doing at all, it is not worth doing well.] The Krebs cycle is not only a way of producing lots of ATP but is also a way of producing substances for synthesis. For example, α-ketoglutaric acid can be converted to glutamic acid, an amino acid that is also the starting point for the synthesis of many other amino acids; fumaric acid and oxaloacetic acid are sources for making aspartic acid, and succinic acid is needed to synthesize, among other things, the porphyrin ring that is part of the heme molecule--the component of hemoglobin, myoglobin, and cytochromes. A cell like a bacterium growing in a medium that lacks one or more amino acids must (if it can) make its own, using the Krebs cycle as a source of precursors.

Replenishing reactions

But what happens when a cell drains off the cycle's intermediates for purposes of synthesis? Suppose, for example, that all of the α-ketoglutaric acid is used to make amino acids. The cycle stops because no more oxaloacetic acid is available to combine with the acetyl group from acetyl-SCoA. A cycle is dependent on the maintenance of the appropriate levels of all of the intermediates. When intermediates are drained off for other purposes you need a replenishing or anapleurotic reaction. For this cycle, the replenishing reaction in E. coli is

PEP (phosphoenolpyruvic acid) + CO_2 → oxaloacetic acid.

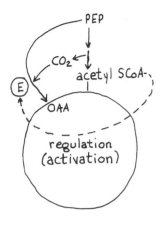

Regulatory enzymes

PEP is the high-energy compound just before pyruvate in the glycolysis pathway (reaction 9). Note that the above reaction can replenish the oxaloacetic acid, but in this process one loses the energy that would be made available by the pyruvate → acetyl-CoA step as well as that derived from the reaction 10 of glycolysis: PEP + ADP → pyruvic acid + ATP. Evidently, having eaten your cake to make amino acids, you have to give up something else.

The replenishing reaction of the Krebs cycle also illustrates the third law of Luria, which states that "nature is almost as clever as biochemists are." The enzyme E that catalyzes the replenishing reaction is a regulatory enzyme. That is, its activity is delicately poised to the needs of the cell. The replenishing reaction is regulated by the concentration of acetyl-CoA: the more acetyl-CoA there is present, the more active the enzyme becomes and the more oxaloacetic acid it makes. Every time you remove some glutamic acid, oxaloacetate goes down, acetyl-CoA accumulates, the replenishing enzyme is activated, oxaloacetic goes up and everything is fine again.

This is an example of enzymes that not only perform their functions, but are regulated in their activity by feed-ahead or feedback mechanism. Notice that acetyl-CoA has no direct relation to the reaction performed by the replenishing enzyme; there is no interaction at the active site of the enzyme. Natural selection has perfected the enzyme protein so that it can combine with the regulatory substance and, by a change in configuration, become more efficient. In other cases, the regulatory substances reduce the efficiency of the regulated enzyme. As discussed on page 55, a regulated enzyme is made up of different subunits. The active site, where the substrate is processed, is in a different subunit from that which combines with the regulator substance. Distortion of the regulatory subunits is transmitted to the functional subunits through the molecular contacts between subunits.

Lecture 9

Anaerobic
respiration

In the last two lectures we have explored a group
of biochemical processes--glycolysis, electron
transport, and Krebs cycle--by which cells extract
energy from organic molecules and store it in the
form of ATP. Such processes are common to most if
not all organisms: glycolysis is present in practically all organisms except a few types of bacteria. Likewise, electron transport is missing
from certain bacteria that cannot synthesize the
porphyrin ring needed for the heme of the cytochromes. These bacteria are anaerobic since they
cannot use oxygen to accept electrons and some
are actually inhibited by oxygen. They must live
by fermentation, and this forces them to live in
special environments.

Even bacteria that have an electron transport
system may be anaerobic if they are unable to
transfer the electrons to oxygen and must use some
other electron acceptor in order to get ATP. This
is called anaerobic respiration. For example,
near the Charles River or in the shallow parts of
Boston Harbor the bottom mud is black and the air
in summer smells foul because of hydrogen sulfide
H_2S. This is because a bacterium (Vibrio desul-
furicans) uses sulfuric acid instead of oxygen as
electron acceptor converting it to H_2S, which can
escape as gas or, if iron salts are present, produce iron sulfide--which is black and the cause
of black mud.

Another example is that of nitrate reduction.
When soil becomes very soggy with water, oxygen
cannot get through fast enough. Then a whole
series of bacteria use nitrate $-NO_3^-$ in the soil
as electron acceptor and generate nitrite $-NO_2^-$
and ultimately nitrogen gas. Since nitrate is
the most valuable nitrogen source for plants,
this bacterial denitrification depletes the soil
and makes it harder for plants to grow. These
two examples of organisms using unusual electron
acceptors--sulfate or nitrate--are not biochemical
oddities. They are simply variations on the
general theme of electron transport to get energy
and must have evolved as adaptive modifications
of the oxygen-using process.

Photosynthesis

Now let us go back one step in evolution. The
Krebs cycle potentially converts all of the carbon
of glucose and other organic substance to carbon
dioxide. Why then does CO_2 constitute only about
0.04 percent of the gases in the earth's atmos-

Calvin cycle

phere? Because it is fixed into organic materials
by a variety of biological processes, the most
important of which is <u>photosynthesis</u>.

Photosynthesis is a remarkable process, which
green plants, algae, and some bacteria use to convert the carbon of CO_2 to organic matter. [Note
that the reaction that replenishes the Krebs
cycle (page 78) also involves fixing CO_2 to organic matter; but it is a small capacity reaction,
not geared for massive fixation and in addition
requires a supply of organic compounds.] Photosynthesis operates by fixing CO_2 through a cycle
of reactions called the Calvin cycle after an
American biochemist (no relation of the 16th
century Protestant reformer). This cycle has
many instructive features.

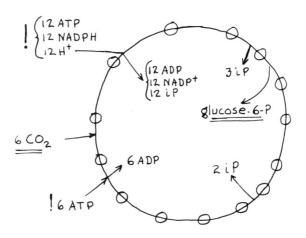

Each circlet means one or more carbon compounds

(Figure represents 6 runnings of the cycle)

The cycle takes in six molecules of CO_2 one at a
time and through a series of steps converts them
into one molecule of glucose. The important
points are:
1. Carbon enters only as CO_2 and exits only as
sugar.
2. To convert 6 moles of CO_2 to one mole of
sugar $C_6H_{12}O_6$ the cycle uses 18 moles of ATP, converting it to ADP and iP. This makes sense: when
a cell burns sugar to form CO_2 in glycolysis and
the Krebs cycle, it releases energy and stores
some of it in the form of ATP. So, to make sugar
from CO_2 there is need to use some ATP.
3. Just as glycolysis and the Krebs cycle release

electrons to NAD^+, the reverse process of making
sugar from CO_2 requires electrons. In other
words, to reduce CO_2 to sugar one needs a source
of <u>reducing power</u>. The cycle provides it in the
form of NADPH (the relative of NADH; remember?).
Twelve moles of NADPH are needed for every mole
of sugar made; therefore the entire reaction
should be written

$$6CO_2 + 12H_2O + 18ATP + 12NADPH + 12\ H^+ \rightarrow$$

$$\text{glucose-6-P} + 6O_2 + 18ADP + 12NADP^+ + 17iP + 6H_2O.$$

NADPH is not a specialty of photosynthesis: it is
used by most cells as the favorite electron donor
for biosynthetic purposes.

[The Calvin cycle is used not only for photosynthesis but for all organisms that must make
their organic carbon from carbon dioxide. Some
important groups of bacteria, for example, oxidize H_2, or H_2S, or S, or even CO (the exhaust
gas) and use the electrons from these compounds
to get NADH or NADPH and to store energy as ATP
made through an electron transport system. They
use CO_2 for carbon source, fixing it by the Calvin
cycle. This, however, contributes very little to
the overall CO_2 fixation on earth.]

Light reactions   Note that in the equation of photosynthesis I
have thrown in H_2O and O_2 and have left out light.
The Calvin cycle constitutes the so-called <u>dark
reaction</u> of photosynthesis. What does light do?
It provides the energy needed to make ATP and
NADPH by the <u>light-reaction</u> part of the process.

In plant cells this takes place in a complex
apparatus called <u>chloroplast</u>.

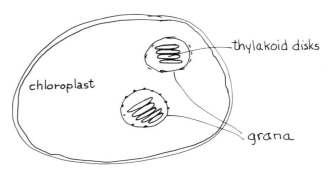

Chloroplasts contain pigments that absorb specific
wavelengths of light and convert it into chemical
energy. Plants are green because the chloroplast

pigments, <u>chlorophyll</u> and <u>carotenoids</u>, absorb the
red light. In the chloroplasts the molecules of
chlorophyll and carotenoid are present in membrane
sacs, called <u>thylakoids</u>, which are stacked in the
so-called <u>grana</u>. The arrangement is such that
the excitations produced in the pigment molecules
by the quanta of light are transferred to a
special group of chlorophyll molecules (pigment
system I or PSI) which act like a condenser accumulating excitation energy. Then electrons are

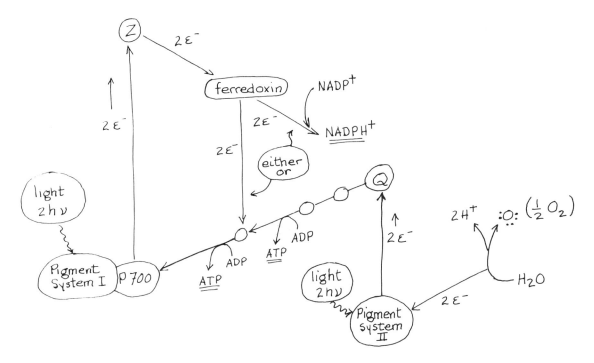

pulled out and transferred to molecules of <u>ferre-</u>
<u>doxin</u>: a small protein that is literally loaded
with iron atoms. The iron atoms of ferredoxin
that receive electrons from chlorophyll are then
in the reduced Fe^{++} state and are powerful donors
of electrons.

   These electrons can then be used in a variety
of ways. Some are returned to chlorophyll through
a series of reactions that produce a substantial
amount of ATP. Other electrons from reduced
ferredoxin are side-tracked to produce NADPH.

   At this point we may have generated all the
ATP and NADPH we need to fix CO_2 to form sugar in
the Calvin cycle. Naively, we might believe that
everything is settled. But once more the first
law of Luria raises its ugly head. When electrons

are taken from reduced ferredoxin to make NADPH,
they cannot be returned to chlorophyll. To regenerate chlorophyll we need a replenishing trick.
In plant cells, the trick is to use a second set
of pigment molecules, PSII, which when excited by
light also release electrons like PSI. These
electrons are transferred to the chlorophyll
molecules of PSI. But the difference is that the
chlorophyll molecules of PSII can regain their
electrons from water, releasing oxygen:

$$2 \ H_2O \rightarrow O_2 + 4 \ H^+ + 4 \ e^- \ .$$

This is why plant photosynthesis generates O_2.
This reaction is the source of all the oxygen in
the atmosphere. The cells are happy, having
balanced their equation, and we can take a deep
breath. (Note that H_2O cannot provide electrons
directly to PSI.)
     Experiments with $H_2{}^{18}O$ have shown that all
oxygen atoms come from water, not from CO_2. That
is why the reaction on page 85 was written that wa
way. The action spectrum, that is, the relative
efficiency of various wavelengths, for plant
photosynthesis looks like this:

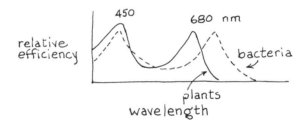

There are two absorption maxima, one in the red
at about 680 nm, the other in the blue at about
450 nm. The red maximum is also the one that is
efficient for photosynthesis. Chlorophyll, a
green substance, has a prophyrin ring similar to
the ones present in the heme of the cytochromes,
which are yellow to red.
     Plants are not the only organisms that carry
out photosynthesis: some bacteria do it too, but
they do not generate O_2 because they take electrons from substances like H_2S or organic compounds rather than from water.
     Since photosynthetic bacteria do not produce
oxygen, their photosynthesis is inhibited by air.
If you look at a pond in the summer, you find

green algae spread on the surface of the water.
The photosynthetic bacteria locate themselves
below the algae, where there is little oxygen
left. The light that reaches them has passed
through a layer of algae, but they can still use
it. The bacterial pigments prefer the longer
wavelengths of light, which are not absorbed by
algae.

How efficient is photosynthesis? In most
plants 1 to 3 percent of the energy absorbed as
light is converted into sugar. In sugar cane the
yield can be as high as 8 percent; this is why
sugar cane is such a fantastically good crop (except, of course, that cutting it is such a back-
breaking work!).

Carbon cycle

Let us now follow what happens in nature to a
few interesting atoms. Consider first carbon:

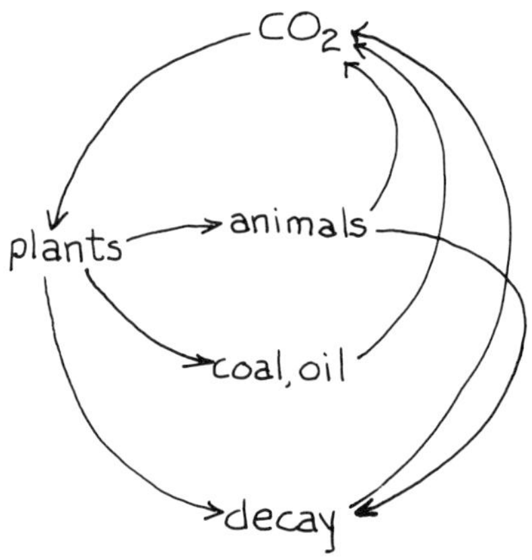

The CO_2 in the air enters the organic cycle mainly
through photosynthesis. The carbon in plants
either (1) is eaten by animals, (2) decays when
plant material dies, or (3) is converted to coal,
oil, and natural gas, which all come from fossil
plants. From animals the carbon can go to decay
(dead animals, like dead plants, are digested by
molds, bacteria, etc.) or be released as CO_2, the
by-product of respiration. Plants also respire
and release carbon as CO_2. Finally, coal, oil,
and gas can be burned to produce CO_2. Thus carbon
in nature is continuously recycled and the concentration of CO_2 in the atmosphere remains nearly
constant. The exceptions are coal, gas, and oil,
which are burned faster than they are made.

Nitrogen cycle

Next, let us see what happens to nitrogen:

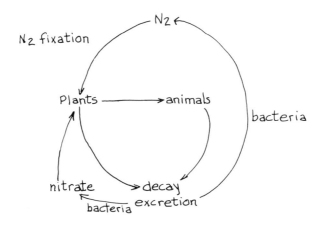

Nitrogen as N_2 makes up 78 percent of air. Some
bacteria, as we have seen, convert nitrate to N_2.
There must, therefore be a process for converting
nitrogen to organic substances usable by organisms.
This process, nitrogen fixation, is carried out by
bacteria, algae and by some plants (legumes such
as alfalfa and beans) in symbiosis with bacteria
in their root nodules. Nitrogen fixation, like
CO_2 fixation, requires ATP and NADPH and converts
N_2 to ammonia, which then enters into plant material. A variety of processes release nitrogen
from plant and animal decay in various forms,
which in the soil can be reutilized by plants or
converted to N_2.

Comparable cycles exist for other elements that
are used by animals. For example, if there were
no sulfur cycle, we would not have the sulfur
needed for proteins. The basis of all these cycles
is the formation of organic compounds from inorganic ones and the subsequent interconversion of the
various organic compounds to make the substances
of the cells according to the program of the genes.

Further reading   Lehninger's Chapter 27, "Photosynthetic Electron Transport and Phosphorylation," is an appro-
priate reading supplement.

Lecture 10

Biosynthesis

In the last several lectures you have seen how organisms obtain energy. In very general terms, we may say that for any substance available on earth from which it is possible to extract some energy by oxidation there is some kind of organism that has found a way to do so.

Energy is first stored as ATP and then used to make possible the reactions by which substances derived from food are built into specific macro-molecules of the cell. How this building is done is our next topic.

Energy from ATP is also used to accumulate substances from the medium in the cell, to convert one substance into another, and to join together smaller molecules into larger ones.

We have already learned that water passes through cell membrane rather quickly. Electrically uncharged substances such as glucose can pass through the membrane, but only rather slowly, the rate depending on their chemical structure. Ionized substances do not pass through membranes unless the cell has specific transport systems to ferry them across. A transport system that works against a concentration gradient, that is, carries molecules from a region of lower concentration to one of higher concentration, requires energy and is called <u>active transport</u>:

$$\Delta G \text{ for active transport} = RT \ln \frac{[\text{inside}]}{[\text{outside}]}$$

For example, in bacterial cells the concentration of K^+ may be 1,000 times higher than in the medium and is maintained by active transport. In the cells of the human body the concentration of potassium is 50 to 100 times higher than in the blood plasma. Active transport requires energy, either from ATP or from other sources.

Synthesis of amino acids

Next, how is ATP used in interconversion of substances? Consider the synthesis of the amino acid proline by <u>E. coli</u>. Proline is formed in several steps from glutamic acid.

E. coli, which can grow using only glucose and
ammonium ions as sources of carbon and nitrogen,
must first make glutamic acid. The appropriate
Krebs cycle intermediate picks up an amino group
and forms glutamic acid. Then some glutamic acid
is converted into proline. Look at the first step.
Do you see something familiar? In this step an
acid -COOH is reduced to an aldehyde -CHO. This
requires two electrons (from NADPH) and energy
(from ATP). The aldehyde, being reactive as all
aldehydes are, reacts with the amino group on the
same molecule, closing a ring to form proline.
(Remember the formation of the closed ring form
of sugars! There the aldehyde reacted with an
alcohol group rather than an amino group.)

Feedback regulation

Each step in a biosynthetic pathway is catalyzed
by a specific enzyme. For example, the first step
in the synthesis of proline from glutamic acid is
catalyzed by a dehydrogenase that adds two electrons to a carboxyl group. There exist other
enzymes that catalyze similar reactions on other
acids but not on glutamic acid. This is a crucial-
ly important fact--enzymes are specific not only
for the kind of chemical reactions they catalyze,
but for a specific substrate. Because of this
specificity, the activity of individual enzymes
can be modulated according to the needs of the
cell. Let us see how.

A cell needs both glutamic acid and proline,
as well as many other things, to make protein. If
the synthesis of any one amino acid were allowed
to proceed unregulated, it might produce unbalance.
The quickest regulation is by means of a direct
feedback mechanism: the level of proline regulates
the pathway by inhibiting the first enzyme in this
pathway. The inhibition is specific and can actually be demonstrated by adding proline to the purified enzyme: its specific activity decreases.
When in the living cell the level of proline increases, proline inhibits the enzyme and less proline is made; when there is less proline available,
the inhibition is released and more proline is
made. This simple negative feedback loop is analogous to the feed-ahead regulation of the replenishing reaction for the Krebs cycle.

The synthesis of arginine, another amino acid,
illustrates similar principles. Here again the
essential steps involve the use of a high-energy

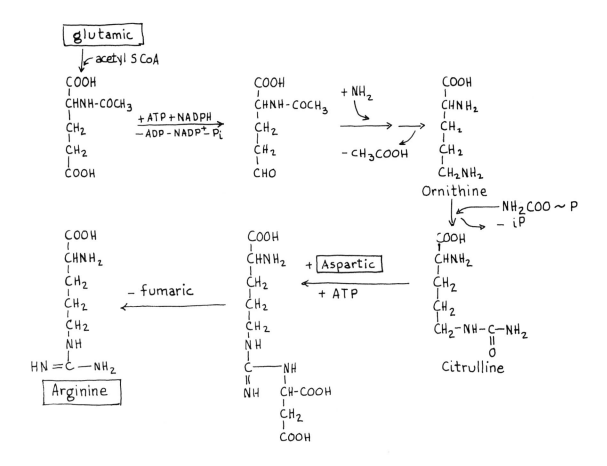

compound such as ATP and the use of an electron donor such as NADPH in order to permit reduction. In addition, this pathway involves the condensation of two compounds, citrulline and aspartic acid, which is made possible by the use of a molecule of ATP.

One more example: threonine, another amino acid synthesized from aspartic acid by a series of reactions. As we would expect by now, this pathway is regulated, with threonine inhibiting the first enzyme:

In this case, however, there is a complication.
Threonine is itself a precursor of another essential amino acid, isoleucine. Again, high levels
of isoleucine inhibit the first enzyme beyond
threonine, that is, the first enzyme on the specific pathway that leads to isoleucine. Try to
figure out what would happen if isoleucine inhibited the first enzyme after aspartic acid instead
of the first enzyme after threonine.

There are other complications. For example,
the conversion of aspartic acid to aspartic semi-
aldehyde is the first step in several pathways
leading to different substances. Complications
are avoided in a clever way: the E. coli cell
has not one, but several enzymes for catalyzing
the same reaction, and each enzyme is sensitive
to inhibition by just one of the end products.
This results in an accurate modulation of the
amount of products rather than a full inhibition.

One important caution: the reason that regulatory feedbacks work promptly and precisely is that
in the cellular pool the small-size molecules such
as amino acids are present in very small concentrations, determined by the relative rates of biosynthesis and of utilization. For example, most
amino acids in E. coli are present in amounts that
would be used up in 30 to 60 seconds of growth.
If an amino acid stopped being made, in less than
a minute its level would go to zero. Because of
this, even slight changes in the rate of protein
synthesis change the levels of amino acids and
trigger the feedback control.

These examples illustrate the regulation at
the level of enzyme function. There is another
form of regulation that takes place at the level
of enzyme synthesis. Addition of arginine, for
example, not only inhibits the first enzyme in its
pathway but also reduces the synthesis of all the
enzymes of the pathway. We shall learn about this
second type of regulation, called repression or
induction when we discuss control mechanisms of
gene action.

Synthesis of   The next level of biosynthesis is that of the
macromolecules   coded macromolecules. The situation here is different from the synthesis of amino acids or of
sugar polymers. To make proline from glutamic
acid, for example, all that was required was three
enzymes plus glutamic acid, ATP, and NADPH. Likewise, to make glycogen from glucose only a few

enzymes are needed, plus ATP or an equivalent. In order to synthesize DNA, the appropriate enzymes plus the properly activated nucleotides are not enough. There must also be DNA to act as the tape, the template to direct the synthesis of the new DNA. Likewise, in order to synthesize protein, in addition to amino acids attached to the proper carriers and various enzymes, coenzymes, and other factors, a molecule of RNA is needed to serve as a template. A mixture of all the precursors and enzymes will not make the coded substances in the absence of the appropriate templates.

Templates

The template does not provide energy, or chemical reactivity, or pieces for synthesis; it provides <u>information about the order</u> in which the pieces are to be assembled. [It may seem that the template, a piece of DNA or RNA with bases in a specific order, may provide a certain amount of <u>negative entropy</u>, that is, of order. But this is not thermodynamic order: any sequence of 30 nucleotides or 30 amino acids has as much chemical and physical order as any other. What a template provides is biological order, that is, the selection of certain useful sequences among all the possible ones. In reality, however, the enzyme machinery itself has evolved in such ways that without templates the syntheses would proceed very slowly if at all.]

DNA synthesis

What do we know about DNA synthesis? Not yet enough, and new facts continue to turn up to disprove earlier theories.

<u>In vivo</u> synthesis

We should first distinguish between synthesis of DNA <u>in vitro</u> and synthesis of DNA <u>in vivo</u>, in the living cell. How can we measure <u>in vivo</u> synthesis? The most convenient way is to label newly made DNA specifically by feeding to the cell some radioactive thymine or its nucleoside thymidine. Thymine is the only base that is present only in DNA and not in RNA. Thymine is an especially good label because it cannot be converted to uracil or cytosine in the cell: the methyl group on thymine cannot be removed without breaking the ring and destroying it. Thymine can be labeled with ^{3}H or ^{14}C in specific positions and traced by the β-rays emitted in radioactive decay.

Radioisotopes

[^{14}C and ^{3}H have different radioactive properties. When an atom of ^{14}C disintegrates, it emits an electron with 0.16 MeV (megaelectron volts) of energy. In water or in photographic emulsion this

electron has a path of one or two millimeters.
(The mean free path of an electron in a given
medium is proportional to the energy with which
the electron is emitted from a radioactive atom
when its nucleus disintegrates.) When tritium
decays, it releases electrons of much lower energy,
about 0.015 MeV, which can only move a fraction of
a micron in water. Thus ^3H is useful for localized
autoradiography: a cell or a fiber of DNA labeled
with ^3H-thymidine, placed on a glass slide and then
covered with photographic emulsion, releases electrons, which alter the sensitive grains in the
emulsion only near the site they come from.
Double- and single-stranded DNA, for example, can
be distinguished because double-stranded DNA has
twice as much label and generates a line of silver
grains twice as heavy in a radioautograph.

. . . . single stranded
.... double stranded

To measure total amounts of material synthesized
by incorporation of label, one can use ^{14}C or ^3H
and count the number of disintegrations per unit
time. Besides ^3H and ^{14}C, one can use ^{32}P supplied as iP (1.7 MeV electrons). Commercial
counters have built-in computers that can analyze
counts from ^3H, ^{14}C, and ^{32}P in a mixture of all
three.]

Density label   Instead of using radioactive thymine to label
DNA one can use a thymine analog that fools the
cells and sneaks into DNA. An interesting analog
is bromouracil or BU, which contains a bromine atom
in place of the methyl group. Bromine is about
the same size as the methyl group but is much
heavier, so that DNA with BU in place of thymine
is heavier than normal DNA. (BU also makes more
pairing mistakes during DNA replication: mutations result when BU pairs with a guanine instead
of adenine, because a G-C pair replaces an A-T
pair. But this is another story.)
     One more way of density labeling DNA and other
macromolecules in bacteria is by growing them in
a medium that has as nitrogen source ammonium
salts -$^{15}NH_4$ instead of -$^{14}NH_4$. This causes an
increase in the molecular weight, for example of
DNA, by one unit whenever a ^{15}N is incorporated
instead of ^{14}N.
     Whether the density label is BU or ^{15}N, the ex-

periments involve the same principle: measurement of the rate at which molecules sediment in an ultracentrifuge within a solution whose density varies over the length of the tube.

How is DNA replicated _in vivo_? We already know the blackboard answer: the helix unwinds and two new complementary strands are formed. Each of these remains associated with one of the old ones.

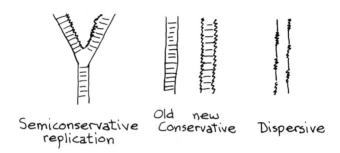

Semiconservative   Old   new   Dispersive
replication   Conservative

This kind of replication is called semiconserva- tive (not because of its being half-way between the John Birch Society and the Democratic Party): each new DNA molecule that is formed is half old and half new. Other possible mechanisms would include conservative replication, which would generate one completely old and one completely new molecule; and dispersive replication, in which multiple breakage and reunion results in DNA strands with both old and new pieces. Each of these types of replication gives rise to speci- fic predictions as to how the density label pres- ent during one cycle of DNA replication would be distributed. This, in turn, is reflected in how the DNA molecules or their fragments would sedi- ment in a density gradient centrifugation experi- ment.

Semiconservative
replication

Semiconservative replication was proved to be the correct mechanism, at least for bacteria, by Meselson and Stahl in 1957. They grew bacteria in a ^{15}N medium for several generations so that all of the macromolecules, including DNA, were uniformly labeled with ^{15}N. Then they shifted the bacteria to a ^{14}N medium. At various times they took samples, extracted the DNA, broke it into short pieces, and measured its density in a CsCl gradient. [When a solution of CsCl is centri- fuged for several hours, the heavy cesium ions tend to concentrate at the bottom of the tube, establishing a gradient of density, for example,

ρ = 1.8 at the bottom of the tube, 1.6 at the top.
Molecules centrifuged with the CsCl become distributed at the levels corresponding to their respective densities.] DNA with ^{14}N or ^{15}N can be readily separated. E. coli DNA grown with ^{14}N medium
has a density of about 1.7, DNA with ^{15}N about
1.73. This is quite a large difference, easily
measured by centrifugation.

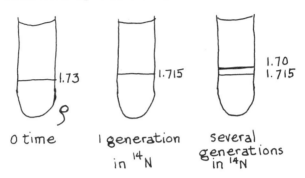

O time   I generation   several
                in ^{14}N   generations
                                       in ^{14}N

One generation after the shift from ^{15}N to ^{14}N
there was no DNA left with ρ = 1.73; all DNA was
in one band at ρ = 1.715 (hybrid density). One
generation later there appeared a band at ρ = 1.70,
which increased in amount with each generation,
but the band at ρ = 1.715 remained present in the
same constant amount.

What does this mean? First, since at the end
of one generation there were no more pieces of
DNA with density 1.73, the conservative mechanism
was excluded. The hybrid density pieces must contain one strand of light and one strand of heavy
DNA. At the end of a second growth cycle, each
hybrid molecule has formed two molecules, one hybrid and one totally light. From then on the
amount of hybrid DNA present must remain constant
if there is no destruction of strands, while the
amount of light DNA pieces increases since all the
new DNA strands being made are light. The dispersive replication mechanism is also eliminated
because the bands remain sharp and do not overlap.

These results also imply, therefore, that in
the growth of E. coli there is no significant
destruction of DNA. This turns out to be true of
all cells. DNA is chemically the most stable compound in the organism. In a multicellular organism
there is some loss of DNA, of course, due to cell
death.

The Meselson-Stahl experiment also suggests

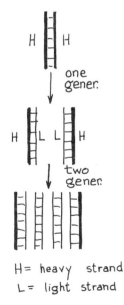

H = heavy strand
L = light strand

Circular DNA

that the synthesis of DNA is <u>sequential</u> and
<u>directional</u>. The best evidence of this was pro-
duced by the radioautographic experiments of John
Cairns. Cairns labeled bacteria with ^{3}H-thymine
for various times; then he extracted the DNA very
gently and looked for the patterns of labeled DNA
by autoradiography. For short periods of time,
increasing periods of labeling resulted in increasing lengths of labeled DNA. If the cells were
labeled for times at least equal to the doubling
time of <u>E. coli</u>, fully labeled <u>circles</u> of DNA began to be seen. The DNA molecules in bacteria,
Cairns concluded, must be circular.

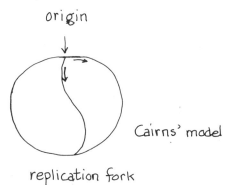

By more complicate experiments, Cairns showed
that the DNA circles always replicate beginning
at the same point of origin. He obtained figures
that suggested a circle replicated part way from
one point in one direction. More recent experiments by Huberman and others have shown that the
replication does not begin at one of the forks,
but at a point midway between them and then proceeds in both directions.

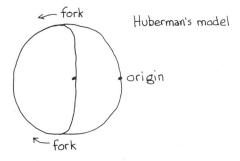

Note that the same image results with either uni-
directional or bidirectional modes of replication.

Bidirectional replication was supported by more elaborate autoradiographic experiments.

The replication of a single DNA molecules as long as the entire genetic complement of E. coli, about 1 mm of DNA, is sequential starting from one point and proceeding in both directions until the two replication forks meet and two separate circles are produced.

The existence of circular DNA in bacteria gave rise to a whole series of more complicated models for the replication of DNA. At the moment, however, there is no certain model of how DNA molecules are made, of how they separate, or of how they are distributed to daughter cells or chromosomes.

In vitro synthesis

What about DNA synthesis in vitro? Note that the above experiments say nothing about the bio-chemistry, that is, the enzymes involved in DNA replication. These can only be studied in vitro, using cell extracts.

DNA polymerases

One enzyme, called DNA polymerase I (Pol I) can make DNA in a test tube by semiconservative repli-cation of a DNA template if it is given the four necessary nucleotides. These must be present in the activated form as 5' deoxytriphosphates (dATP, dGTP, dCTP, and DTTP). The polymerase I joins

together the nucleotides releasing iPP. This en-zyme was discovered by Arthur Kornberg in 1955 and was believed to be the enzyme that made DNA in the cell until 1969, when Cairns discovered some E. coli mutants that lacked Pol I activity but made DNA perfectly well! The ensuing search led to the discovery of two more DNA polymerases, Pol II and Pol III, in E. coli cells. Again, mutants lacking Pol II could still make DNA, so that Pol II, like Pol I, was not essential. Poly-merase III, however, seems to be the real McCoy. Mutants (called dna E) with a Pol III enzyme that is temperature sensitive (active at 25°C but inactive at 40°C) grow at 25°C or 35°C but can make no DNA at 40°C.

Irrespective of which enzyme is the one and
only DNA polymerase, the important thing is that
all these enzymes work by the Kornberg mechanism,
using only 5' deoxytriphosphates as precursors,
removing two high-energy phosphates in the form
of pyrophosphate iPP, which is then split to iP,
and linking the nucleotide to the 3' hydroxyl
group of the preceding nucleotide in the growing
DNA chain. Each nucleotide added is one more link

in the 5'=3'-phosphate chain and provides a new
free 3' hydroxyl for further addition. All
enzymes can only add nucleotides from the 5' to
the 3' direction. What the template DNA does is
to line up the proper bases for base pairing.
What the enzyme does is to catalyze the joining
reaction.

What separates the strands of the double helix
during replication? No one knows for sure. Possibly the polymerase molecule causes the separation by pushing the strands apart (after all, each
base pair is held together by only about 6 to 9
kcal/mole of bond energy). It seems more likely,
however, that other specific proteins perform the
separating function.

Further reading   Watson's Chapter 9, "The Replication of DNA,"
goes much beyond the scope of this lecture. Stent's
Chapter 9, "DNA Replication," will help understand
the methods used in this area of research.

Lecture 11

All synthesis of nucleic acids, whether DNA or
RNA, consists of the same overall chemical process,
in which nucleoside triphosphates serve as donors
of nucleotides to build chains in the 5'-3' direction. The products of DNA or RNA synthesis consist of polynucleotide chains, which can be written in two ways:

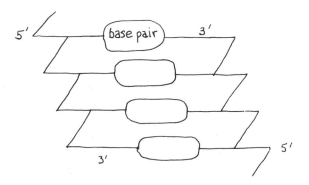

Each strand of DNA has a free 5' end (with or without one, or two, or three phosphates) at one end
and a free 3' hydroxyl end. Because of the nonsym-
metrical arrangement each strand has a polarity,
either 5' → 3' or 3' → 5'. In a DNA molecule the
two strands have opposite polarities:

The bases in one strand are upside-down with respect to the bases in the complementary strand.
The only way two nucleotide strands can come together and form a double helix à la Watson and
Crick is in this antiparallel way. There is no
way to build them together to form a double helix

with parallel strands. More generally, whenever
there is pairing between two nucleic acid strands
(DNA-DNA, DNA-RNA, or RNA-RNA), it is always anti-
parallel. Thus there are two rules of pairing:
(1) the normal base pairing always occurs between
G and C and between A and T (or between A and U
in RNA); and (2) the two strands are antiparallel.

   This creates a curious difficulty. Both
strands of a double-helix DNA are replicated in
the same direction. Yet enzymes make DNA strands
only in the direction 5' → 3'. How then can they
replicate both strands? The 3' → 5' strand is
easy to replicate, but how can the enzyme replicate the 5' → 3' strand, which would require
building in the 3' → 5' direction?

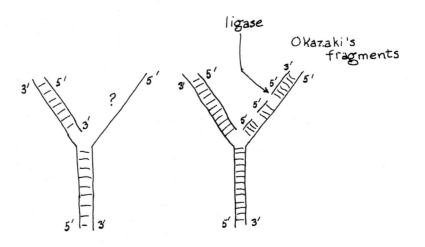

A biochemist named Okazaki made a helpful discovery. By labeling a new DNA for only a few
seconds, he found that the label was in very short
pieces of DNA. These then disappear and are incorporated into longer pieces of DNA. So he
proposed that synthesis of the 3' → 5' chain
occurs through a backward formation of pieces in
the 5' → 3' direction, which are then joined together. This joining could be done by a well-
known enzyme, DNA ligase, which can do just that.
In other terms, Okazaki's model is that one DNA
strand is being made backwards. The Okazaki
pieces may be the real way DNA is made; but this
is not yet certain.

   The problem with all these experiments is that
DNA synthesis is a very fast process (1,000 nucleotides added per second at 30°C). The whole DNA
of a cell like E. coli, over a million nucleotide

pairs long, can be replicated within 30 or 40 minutes.

What happens when the two strands in an intact double helical DNA molecule are separated but not replicated (for example, in making RNA)? The two strands can go back together to reform the original structure by a process called annealing. In the test tube, you can separate DNA strands by setting an alkaline pH or by heating. If you heat DNA in a salt solution to around 90°C, the strands separate because the thermal energy breaks the hydrogen bonds that held them together. If the solution is then cooled slowly down to room temperature, the correct double helical structure reforms. If instead the solution is cooled by quickly immersing it in ice, the double-stranded structure cannot reform because there has not been time for the trial-and-error collisions needed to regenerate the correct double strand structure. Small local regions of base-pairing may form within each strand (always in an anti-parallel way), but the strands will have the overall structure of a random coil.

There are many ways to determine whether or not DNA has reannealed. Single-stranded DNA can be distinguished from double-stranded DNA by centrifugation, since the former, all balled up, sediments faster than the latter (more rigid, more elongated, subject to more friction). More convenient are methods based on the different behavior of single- and double-stranded nucleic acids in passing through filter membranes. Still another way is to use the electron microscope: a single-stranded DNA piece looks like a roundish mass, and a double strand looks like a string or thread. Also, there are enzymes that digest single-stranded but not double-stranded DNA and can be used to chew off single-stranded ends from a DNA duplex.

One-strand transcription

Sometimes, for example for some bacteriophages, the two strands of DNA, once separated, turn out to have different densities and can be physically separated: as someone put it, you can get Watson away from Crick. (The reason is that DNA density depends on the content of G, which is denser than the other bases.) This observation is important because it can answer the question, Which strand of DNA is the template for the corresponding RNA?

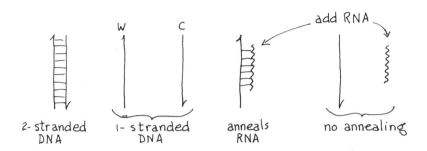

2-stranded DNA   1-stranded DNA   anneals RNA   no annealing

If you have some labeled RNA known to be made by
a given piece of DNA, you can test the ability of
the labeled RNA to anneal with each of the separated DNA strands. The RNA anneals only with the
DNA strand to which it is complementary, that is,
the strand whose transcript it is. The general
conclusion is that, whenever RNA is produced it
is transcribed off only one of the two DNA strands.
For example, in the E. coli chromosome, RNA may be
transcribed from one strand in some places, the
other strand in other places, but never off both
strands at the same site.

Is this situation reasonable? Could evolution
have produced double, or W-C, genes, in which one
DNA strand is coded for one product and the complementary strand, read in the opposite direction,
coded for a different polypeptide product? There
is no a priori reason against it except for the
problem of mutations. If both strands carried
genetic information, every mutation would result
in a change in two genes. This is an enormously
unfavorable situation, sufficient in itself to
discourage the evolution of W-C genes. Since
most mutations are unfavorable, evolution must
have preferred to waste all of the information in
the unused segments of DNA strands rather than
risk the chance of being messed up genetically.

I mentioned earlier the ligase enzyme, which
is present in every cell. It might be there just
to please Okazaki. But it has at least one other
important function. The ligase is part of a group
of enzymes that repairs DNA whenever it is damaged,
for example, by ultraviolet light or by x rays or
by mistakes in replication. Every normal living
cell has such a system for repairing DNA. For
example, ultraviolet light can cause two adjacent
thymine bases in a DNA strand to join in a way
that would prevent DNA from replicating. This is
called a thymine dimer. The repair system starts
working in a series of steps:

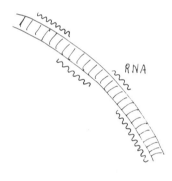

RNA

DNA repair

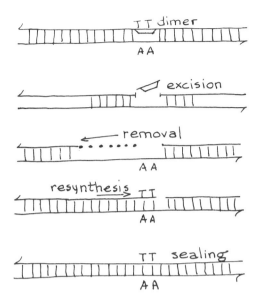

1. an enzyme breaks the damaged strand of DNA on
either side of the joined bases;
2. another enzyme chews up a small piece of the
DNA on one side of the damaged section;
3. a third enzyme, polymerase I, rebuilds the
missing piece again using the opposite strand as
a template;
4. the ligase seals the break.

There is a human genetic disease called xero-
derma; the affected people lack one of these
repair enzymes. They can live well, but if they
get exposed to too much sunlight their skin cells
are easily damaged and tend to produce cancers.
The defect is in the repair system, which works
badly. Its mistakes can lead to cancer by mutating
one or more genes.

RNA synthesis   Let us now consider the synthesis of RNA. RNA
is made from riboside triphosphates, usually by
base pairing with one strand of DNA. The mechanism
is conservative since the RNA comes off and the
DNA strands return together. The RNA made in any
cell falls into three classes:
1. Ribosomal RNA. A ribosome, part of the
machine to make protein, consists of two subunits.
Each subunit contains one molecule of RNA, always
the same in a given organism. Bacterial ribosomes
differ somewhat in size from those of animal cells.
2. Transfer RNA or amino acid adapters. These
molecules are 70 to 85 nucleotides long and are
specific for each individual amino acid. They

serve to insert the amino acids into proteins.
3.   Messenger RNA. These are of variable size
because they represent the transcript of individual
genes or groups of genes, which will then direct
the synthesis of proteins.
Thus all species of RNA participate in one way or
another in protein synthesis.

The overall process of RNA transcription from
DNA and its translation into protein in bacteria
has been visualized by Oscar Miller in some superb
electron microscopic pictures which look something
like this:

Polysomes

DNA

This is the appearance of nascent strands of E.
coli RNA still being made on template DNA. The
long central fiber is the DNA and the fibers
coming off it are the partially synthesized RNA
molecules. The small particles visible on all
of the RNA fibers are ribosomes. As soon as the
messenger is even partially available, the ribosomes attach themselves to it and begin making
protein. One molecule of mRNA may have up to 10
or 20 ribosomes attached at a given time, each
making a polypeptide chain. Sometimes, however,
there are no ribosomes attached to the nascent
RNA strands. These RNA molecules do not code for
protein, that is, they are either tRNA or rRNA.

The picture of RNA synthesis in animal cells is
different: ribosomes are not visible on nascent
RNA. This is because in eucaryotic cells messenger RNA picks up ribosomes and makes proteins only
after it passes from the nucleus to the cytoplasm.

RNA

DNA

RNA polymerase

In every cell there is at least one enzyme,
called RNA polymerase, that catalyzes the synthesis of RNA off of the DNA in the usual 5' → 3'
direction. Molecules of RNA polymerases may be

seen in Miller's electron micrographs where RNA
is still attached to DNA. This enzyme must be
able to recognize the <u>starting signals</u> that indicate the beginning points of sections of the DNA
that are transcribed into RNA molecules. RNA
polymerase is the key enzyme for the expression
of gene function. Genes are stretches of DNA;
the polymerase recognizes where to begin and
where to end on the DNA to make the <u>RNA transcript</u>
of a gene or a group of genes.

The above picture of nucleic acid synthesis is
fairly straightforward. By the intervention of
one or a few enzymes, DNA is replicated to generate
more DNA. By the action of RNA polymerase DNA
segments are transcribed into RNA molecules.

Viruses

There are some organisms, however, that include
special cases with respect to the nature and replication of their genetic materials. These organisms
are the <u>viruses</u>. Viruses are not cells; they must
use living cells of other organisms to reproduce.
Viruses, when in the free state, are in the form
of particles or <u>virions</u> that consist of a piece of
genetic material--a nucleic acid molecule--plus
one or more protein coats. All viruses have some
kind of mechanism for inserting their nucleic acid
into specific host cells, where it is replicated
and more virus is made. Viruses that grow in
bacteria are called <u>bacteriophages</u> or <u>phages</u>.

Some viruses contain double-stranded DNA as
genetic molecules; in principle, therefore, they
present no difficulty. Their DNA can be replicated
as usual using either the cell's machinery or sometimes by using a special viral machinery.

Some viruses, however, contain odd types of
nucleic acid molecules. One class has a circular
single-stranded DNA. When this enters a cell, an
enzyme present in the cell, probably DNA polymerase I, makes a complementary strand so that the
single-strand circle becomes a double-strand circle,
which then proceeds to reproduce in the 2-strand
form. After a while some circles start making

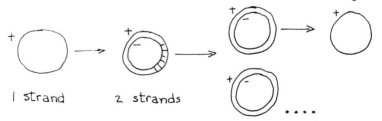

1 strand   2 strands

replication

single-stranded DNA circles, which enter into complete viruses. You may have guessed that this
mechanism is arranged in such a way that it always
selects the same strand--Watson or Crick--to put
in the complete virus particle. But how this
happens we do not know.

RNA viruses

Other viruses, for example, poliovirus and some
phages, are even odder: their genetic material is
RNA! This raises a major difficulty. Cells, as
far as we know, contain no machinery to make RNA
from RNA or to make DNA from RNA. Some RNA
viruses, such as polio, solve this problem in a
direct way: the entering RNA, acting as mRNA,
makes an enzyme that can copy the viral RNA, first
making a complementary strand and then more viral
RNA strands.

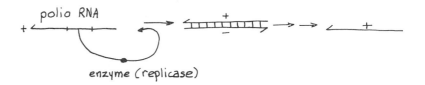

The most complicated and most exciting case is
that of RNA tumor viruses. They are the cause of
many cancers of animals and probably also of man.
Their virions contain, besides single-stranded
RNA, an enzyme called reverse transcriptase, which
can make DNA from RNA! When the RNA of a tumor
virus enters a cell, this enzyme comes in with it
and catalyzes the synthesis of a complementary
DNA strand, thus creating a DNA-RNA hybrid molecule.

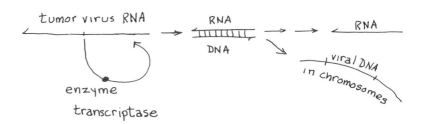

The discovery of reverse transcriptase (in 1970)
was a shock because it had been generally assumed
(without any strong theoretical reasons) that in
macromolecular syntheses information would flow
only from DNA to RNA and from RNA to protein.

[One should not take beliefs too seriously. In
August 1901 the famous philosopher Hegel wrote in
his doctoral dissertation that the planets then
known were of <u>necessity</u> the sum total of the solar
planetary system. Unfortunately, the preceding
January the astronomer Piazzi had discovered
Ceres, and Neptune and Pluto were soon to follow.
As Francis Crick has pointed out, he and Watson
never postulated that information had to flow
only from DNA to RNA rather than from RNA to DNA,
but only that protein should not be able to serve
as template to make nucleic acid. Can you see
why?]

Reverse transcriptase is also important in one
practical respect. It provides tumor viruses
with a mechanism for converting their genetic
material into DNA. This DNA can then remain in
the cell, joining the DNA of the chromosomes. The
permanent presence of viral DNA in cellular chromosomes is believed to be responsible for the transformation of normal cells to cancer cells by
these viruses and possibly for many so-called
spontaneous cancers.

Further reading

Watson's Chapter 11, "The Transcription of RNA
upon DNA Templates," and Stent's Chapter 16, "DNA
Transcription," provide relevant information
beyond what is presented in this lecture.

Lecture 12

Protein synthesis

In the last two lectures we have seen how cells
make DNA and RNA. Now we come to the synthesis
of protein. Again, the cell uses two devices
with which we are already familiar: (1) a set of
enzymes to activate the precursor substances--
the amino acids--and put them together to form
protein chains; and (2) a template. In protein
synthesis, in addition, there is one more problem:
the information of the genes, which is present in
DNA or RNA as a sequence of nucleotides, has to
be translated from the language of the nucleic
acids into the language of the proteins, a sequence
of amino acids. There must, therefore, exist
specific rules of <u>translation</u> and these are called
the <u>genetic code</u>. To decipher the secrets of the
genetic code required a combination of genetic
and biochemical techniques. We shall begin with
the biochemical approaches to protein synthesis.
Then after a discussion of the genetic code, there
will be a gap in the story until we come back to
it with some knowledge of genetics.

Protein synthesis can be measured by monitoring
the incorporation of labeled amino acids into
polypeptides, which are acid insoluble. Is the
protein made in any special place in the cell?
Bacteria are too small for us to localize the
sites of protein synthesis within them. In
eucaryotic cells, protein synthesis is easily
seen to occur in the cytoplasm: a short labeling
with radioactive amino acids followed by autoradiography shows the label localized only on the
cytoplasm. Only later some proteins return to the

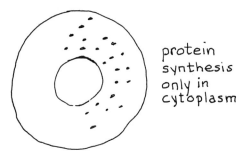

protein
synthesis
only in
cytoplasm

nucleus. Since practically all the DNA in the
cell is located in the nucleus it is clear that
protein is not made directly on or near the DNA.
It is made in the cytoplasm using information
that comes from the DNA via messenger RNA. The

mechanism that transports mRNA from the nucleus where it is made to the cytoplasm where it is used is still unknown. There are indications that the messenger gets processed by enzymes that chew off some parts of it.

Direction of synthesis

Can one tell, from experiments on intact cells how a polypeptide chain is made? As you know, each polypeptide chain contains an amino group at one end and a carboxyl group at the other end. The chains might be made in several ways; for example:

1. in many pieces that are then stuck together;
2. in one piece starting from the carboxyl end;
3. ditto but starting from the amino end; or
4. starting from the middle and growing toward the two ends.

The answer is simple: <u>all polypeptide chains are made starting at the end with the terminal amino group and ending at the terminal carboxyl group.</u>

This was shown in 1960 by Howard Dintzis, who used <u>reticulocytes</u>, that is, immature red blood cells from rabbits. (Mature red blood cells have stopped making protein and cannot be used.) <u>Reticulocytes</u> make one protein only, hemoglobin. This protein, whose properties we shall consider later in some detail, consists of four polypeptide chains, a pair of α chains and a pair of β chains: $\alpha_2\beta_2$. (To each chain there is attached a nonprotein heme group.) The bone marrow of a rabbit can be stimulated to produce reticulocytes in large excess and release them into the bloodstream. In Dintzis' experiments, a reticulocyte suspension was used to label growing hemoglobin chains, which were then isolated from complete hemoglobin molecules. The label was the amino acid leucine, of which there are many copies all along the α and the β chains. Dintzis first found that it took about one minute to make a complete β chain. Then he added to a suspension of reticulocytes the labeled leucine for a short period, about 30 seconds; stopped protein synthesis by chilling the suspension; isolated the complete $\alpha_2\beta_2$ hemoglobin molecules; purified them; separated the α and β chains. Then he broke the β chains into specific

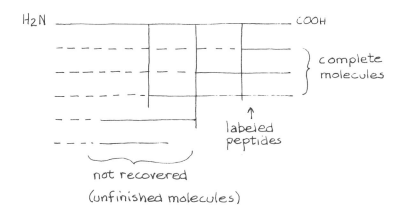

not recovered

(unfinished molecules)

fragments (or peptides) by means of trypsin,
separated the peptides by electrophoresis and
chromatography, and asked how the label was distributed into various regions of the chains.
Where would one expect to find label in <u>finished</u>
<u>hemoglobin chains</u>? Primarily in those portions
of the chains that are made last since the labeling had lasted only one-half the time needed to
make a chain and since what was examined was the
chains that had been finished. What Dintzis found
was that the completed hemoglobin chains were
labeled predominantly near the carboxyl end, and
the closer a fragment was to that end, the more
labeling it had. The peptides near the H_3N^+ end
were unlabeled.

In the reverse experiment labeled leucine was
added for a short period **of** time and then a hun-
dredfold excess **of** nonlabeled leucine was added to
swamp out the radioactive label. Then some time
was allowed to pass before the reaction was
stopped. In this <u>pulse-chase experiment</u>, more
label was found present at the amino terminus
because many partly labeled molecules could be completed in unlabeled medium. [Notice that the key
to interpreting these experiments was that only
finished hemoglobin molecules were analyzed. Unfinished molecules would have contributed labels
in all parts of the chain if they had been examined
along with the finished ones.]

Dintzis' experiments were done in 1960, before
the discovery of techniques for synthesizing proteins <u>in vitro</u>. Since then it has been confirmed
that all protein chains are made in the same way,
starting at the amino end and going toward the
carboxyl end.

Do all proteins start with the same amino acid?
There are reasons to expect that this is so. If
the signal that directs the starting point for
making protein at a particular site on a messenger
RNA were a unique sequence representing a given
amino acid, all proteins in a given organism
might be expected to begin with the same amino
acid. If not, then there must be some other signal that says "start here" but is not itself
translated into protein. (Think of a bus stop
signal that directs a bus to stop 100 feet down
the road.) It seems now fairly certain that all
proteins in all organisms that have been tested
start with the same amino acid, either methionine
or a modified form of methionine called formyl
methionine. Yet, when proteins are extracted from
E. coli cells, methionine is the most frequent but
by no means the only amino acid found at the free
amino end.

The explanation is that, as polypeptide chains
are made, one or more amino acids may be snipped
off one or the other end or even from the middle.
For example, insulin is first made as proinsulin,
a chain of 84 amino acids from which certain
enzymes cut out an internal piece 33 amino acids
long. Only then does insulin become active.

(H₂N) met ———————|- - - - - -| ——————— COOH
         part removed

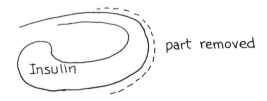
part removed

Insulin

We know that protein chains in general begin
with an amino terminal methionine (or formyl methionine) because we can now study the synthesis of
proteins in the test tube using extracts of bacteria, plant, or animal cells.

In vitro synthesis

The key to success was discovered by Nirenberg
in 1961. Most cell extracts incorporate amino
acids into protein for only a few minutes.

Nirenberg found that, after an extract from <u>E. coli</u> had apparently died, protein synthesis would restart if he added some RNA to it.

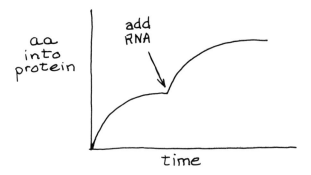

This was an enormously important discovery. It made the study of protein synthesis and its regulation amenable to the methods of enzymology. Biochemists like to work with extracts in test tubes rather than with intact live cells. With Nirenberg's system it was possible to find out what in the extract was needed to make protein. There had to be ribosomes, transfer RNA's (the adapter molecules to which the amino acids must be attached), the amino acids, of course, and the activating enzymes that attach each amino acid to its adapter tRNA. In addition the extract contains some proteins that play various roles in the complicated process of making protein. ATP and GTP are needed to provide energy for specific steps. Finally a messenger RNA has to be present. [The reason extracts from <u>E. coli</u> stopped making protein is that bacterial mRNA is unstable: it is chewed up after it has been used to make a few dozen protein molecules. The mRNA of animal and plant cells lasts much longer. Can you think of some good reason for this stability? Think of the need for rapid adaptation versus stable differentiation.]

    In Nirenberg's extracts, only mRNA served as template. Ribosomal RNA or tRNA did not serve as messengers: they do not code for protein. Instead, the RNA from viruses was a good messenger and directed the synthesis of the same proteins that are found in the corresponding virions. This was a most gratifying result indeed.

    Even more exciting, Nirenberg found that artificially made polynucleotides could serve as messengers, although rather poorly: Each sequence

caused incorporation of specific amino acids:
for example, polyphenylalanine was made under the
direction of poly U. The genetic code was born.

$$pUpUpUpUpUpU = \text{polyuridylic acid}$$
$$\downarrow \qquad\qquad = \text{poly U}$$
$$\text{polyphenylalanine}$$

The necessary existence of rules of translation
between the alphabet of nucleic acids and that of
protein had, of course, been discussed before
Nirenberg's experiments, but only with his system
did it become possible to find out "what meant
what."

Genetic code

    We'll outline now that knowledge although sever-
al ends will be left dangling until we discuss
molecular genetics. The genetic code is a <u>triplet
code</u>: three adjacent nucleotides specify an amino
acid. For example, pUpUpU (or UUU for short) means
polyphenylalanine. Mixed polynucleotides such as
...CUCUCU... in Nirenberg's extracts code for incorporation of mixtures of amino acids into protein.

    The idea of a triplet code is a natural conse-
quence of the fact that 4 nucleotides either one-
by-one or two-by-two ($4^2 = 16$) do not provide
enough combinations to code for 20 amino acids.
Three-by-three ($4^3 = 64$) can work. In fact, it
turns out to be just so, and every one of the 64
possible triplets is used, either as an amino acid
<u>codon</u> or as a signal for termination.

Second Letter

	U	C	A	G	
U	UUU } Phe UUC UUA } Leu UUG	UCU) UCC UCA } Ser UCG	UAU } Tyr UAC UAA [End] UAG [End]	UGU } Cys UGC UGA [End] UGG Tryp	U C A G
C	CUU) CUC } Leu CUA CUG	CCU) CCC } Pro CCA CCG	CAU } His CAC CAA } GluN CAG	CGU) CGC } Arg CGA CGG	U C A G
A	AUU) AUC } Ileu AUA) AUG Met	ACU) ACC } Thr ACA ACG	AAU } AspN AAC AAA } Lys AAG	AGU } Ser AGC AGA } Arg AGG	U C A G
G	GUU) GUC } Val GUA GUG	GCU) GCC } Ala GCA GCG	GAU } Asp GAC GAA } Glu GAG	GGU) GGC } Gly GGA GGG	U C A G

First Letter (left margin)   *Third Letter* (right margin)

But what decides where a triplet starts? Are there "commas"? No. The code is comma-free; there are only start signals, and the reading proceeds down the messenger three nucleotides by threes in a fixed direction from the start point. That is why poly U or poly A, etc., all synthetic nucleotides, are lousy messengers: they lack the start signals.

Universality

The genetic code is universal or nearly so. If you feed to a Nirenberg extract from E. coli an RNA messenger from tobacco mosaic virus or from human blood cells (or vice versa, a messenger from E. coli to a mammalian cell extract) the protein that is made is the same as would be made in the cell of origin of the messenger. This means that the translation dictionary is the same: the codons are the same and the tRNA's recognize the same codons to insert the same amino acids.

Ever since cellular life appeared perhaps 2 billion years ago, nothing much has changed in the genetic code. The enzymes that attach the amino acid to the tRNA must of course have evolved as DNA evolved; but the recognition of an mRNA signal--

a codon--by the tRNA that carries a specific
amino acid has apparently remained constant
throughout the whole period. There are many reasons why this should be so. Imagine a mutation
that changed the recognition of a codon: it
would affect not one protein, but practically all
proteins of the organism. The chances that such
a mutation is not lethal are almost nil. This
conclusion as to the universality and stability
of the genetic code is on a par with the finding
that genetic material in every organism is composed of nucleic acid and functions through its
sequence of nucleotides. It places together all
organisms, past and present, not only in genetic
structure but in mechanism of genetic expression.

The individual codon assignments were decided
on the basis of experiments of many different
types. Certain chemicals can act directly on RNA
and change one nucleotide into another in a chemically defined way. For example, nitrous acid
acting on RNA converts cytosine to uracil (C to U).
If this treatment is used on an RNA that makes a
well known protein--the classical case is the RNA
of tobacco mosaic virus--the altered RNA makes an
altered protein with specific amino acid substitutions. These provide information of codon changes
due to C → U changes.

The most elegant confirmation of the codon
chart came when Nirenberg and his colleagues discovered that a free triplet, for example, the
trinucleotide pApCpC (or ACC), while of course
not causing incorporation of any amino acids into
protein (it has only one codon!) causes the appropriate amino acyl tRNA (aa-tRNA) to stick specifically to ribosomes. If it is added to a mixture
of all aa-tRNA's plus ribosomes which is then filtered, ACC causes only specific threonyl-tRNA to
remain on a filter, while the other amino acids
go right through.

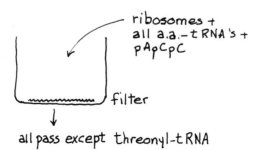

All codon assignemnts have been verified in the
same way. Evidently, a little RNA piece such as
ACC can hold together ribosome and the amino acyl
tRNA strongly enough to keep the latter from going
through the filter. The codon pairs with the
anticodon (for example, UGG if the codon is ACC)
in the tRNA by the usual 2 or 3 hydrogen bonds per
nucleotide.

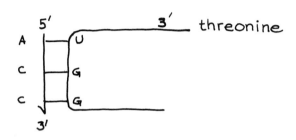

Degeneracy

Of the 64 codons, 61 code for amino acids: the
code is degenerate (that is, it provides more than
one solution for a problem). The extent of
degeneracy varies: for example, there are six
different codons for leucine and only one for
methionine. For each codon on the messenger, of
course, there must be at least one transfer RNA to
bring that amino acid to that position.

Further reading

Watson's Chapters 12, "Involvement of RNA in
Protein Synthesis," and 13, "The Genetic Code," as
well as Stent's Chapter 17, "RNA Translation," are
useful supplements to this lecture and the next
(lecture 13).

Lecture 13

We have seen that the sequence of amino acids in a protein is determined by a series of codons on an mRNA, each codon consisting of three nucleotides. Each codon lines up a tRNA molecule (with its amino acid attached) by base pairing between the codon on the mRNA and the anticodon on the tRNA. Remember that pairing is always between strands of opposite polarity. Each amino acid is specified by one or more different codons, a property we have called degeneracy. But there is also redundancy: for a given codon there is often more than one tRNA. Evolution has provided a safety factor against mutations that would cause disaster if they incapacitated the only tRNA for a given codon. (Keep in mind that tRNA's, like all RNA's, are transcripts of specific DNA sequences.)

Missense mutations

Let us see now what happens when a mutation changes a codon in the gene specifying a sequence of amino acids. If the new codon happens to correspond to a tRNA that inserts the same amino acid, there is no change in the protein. Most commonly, however, the new codon corresponds to a different amino acid. That is, a different amino acid is inserted in the protein. We talk of acceptable missense mutation if the protein with the substituted amino acid is still tolerably functional; that is, the substituted amino acid is similar enough to the one it replaces or at least does not cause too much distortion. Otherwise, the protein fails to fold or if folded is nonfunctional: we call these unacceptable missense mutations.

td mutations

An interesting class of unacceptable missense is temperature-dependent (td) missense: the resulting protein is functional in a permissive temperature range, usually lower than the nonpermissive range. (Can you figure out the reason, from what you know of the bonds that keep together various parts of a protein molecule?) If a totally unacceptable missense mutation happens, the faulty protein may be completely undetectable. Sometimes, however, one can actually find the useless mutant protein: if you have an antiserum that reacts specifically with the corresponding normal protein, you can use it as a reagent to search for the mutant form of the protein.

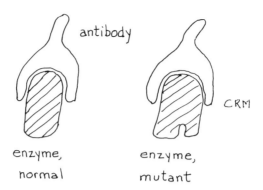

enzyme,   enzyme,
normal   mutant

If you find it and isolate it, you may find out
which amino acid has been changed. A mutant protein detected in this way is called a cross-reac-
ting material or CRM (pronounced crème, as in
crème de menthe). This technique is very useful
in tracing modified proteins in genetics studies.

Termination

You can readily guess that there is another
possibility: a base change in a codon may turn
it into a <u>nonsense codon</u>. There are three nonsense codons UGA, UAA, UAG, which correspond to
no amino acid. They are the <u>termination</u> signals.
When the protein synthesizing machinery encounters
one of these codons synthesis stops. There is no
tRNA corresponding to these codons, so that no
amino acids can be inserted. But the signal is
somehow recognized by the ribosome and causes the
growing polypeptide chain to fall off the ribosome
complex, to detach itself from the tRNA, and to
become free.

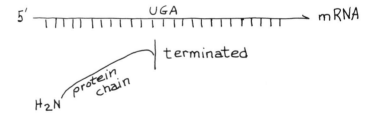

Suppressors

The story of how the nonsense codons were discovered is quite interesting. Since it involved
the study of bacteriophage growth we shall wait a
few lectures before returning to it. For the time
being let us grasp the main principle. Consider
for example a mutation that changes the DNA triplet CTT, which generates mRNA codon GAA, to become
triplet ATT, corresponding to codon UAA on the

mRNA. Since UAA is a termination signal, synthesis of that protein stops, the cell produces only
a fragment of that protein, and if the protein is
a vital enzyme the cell dies--unless...unless the
cell (say, a strain of E. coli) happens to have a
mutant tRNA, which can more-or-less efficiently
put an acceptable amino acid at the place where
the UAA codon exists. This phenomenon is called
suppression of nonsense. A mistaken tRNA corrects
a mistaken codon and saves the cell. [Don't you
wish there were more suppression of nonsense altogether?]

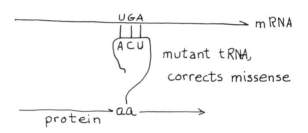

    There is another worry, however. If suppression
requires a mutated RNA, what happens to the codon
that was read by that RNA? The way out is that
the suppressor tRNA's must be, not the major tRNA's
that are used everywhere to put in amino acids, but
some minor, low-grade tRNA's. The cell can afford
mistakes made by these tRNA's in the same way that
people can afford losing their appendixes but not
their hearts.
    Notice some dangers of suppression. How could
a cell survive if an amino acid were inserted whenever a natural termination signal occurs? Proteins would never end. The safety play is that
suppression is usually inefficient, a few percent
at most. It causes little trouble to a cell if 1
percent of its proteins are not terminated properly, whereas even 1 percent of an essential protein made at the site of a mutant gene may keep
the cell going and growing. Practically all bacteria do have some levels of suppression, which
suggests that the mechanism is important in evolution for maintaining and trying out mutated genes.
    Each tRNA molecule must be recognized by the
enzyme that puts the appropriate amino acid on it
and by the codon on mRNA. If a suppressor mutation occurs in the anticodon, the new anticodon
may pair either with the codon for another amino

acid or with a termination codon. Remember that
each tRNA represents the product of a specific
gene.

So much for termination. Let us now see what
happens at the other end of the protein. There
must be an initiation signal, which puts methionine (or formyl methionine) at the start of the
chain. But the code has 61 signals for amino
acids and three termination signals; there is
nothing left for a start signal! In fact, the
start is signalled by one or more special sequences of nucleotides that end with AUG (methionine) but are positioned in a special way on the
ribosome. The "start sequence" is still poorly
understood. One way to try is to use a pure messenger RNA, for example a viral RNA, in a Nirenberg system that lacks all the amino acids except
the starter methionyl-tRNA.

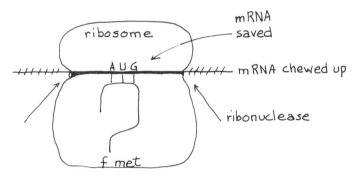

The starting stretch of mRNA becomes trapped in
the groove of a ribosome. Then one adds an
enzyme that digests away free RNA so that only
the part trapped in the ribosome remains and can
be analyzed. It seems that there is not one
start sequence, but a set of start sequences, all
ending with AUG.

The special tRNA for formyl methionine (or for
methionine in animal cells) has the anticodon UAC
corresponding to codon AUG. Why does it not start
a new polypeptide chain every time it finds an AUG
codon for methionine? Because it is a special
tRNA, which works only at the beginning of the
proteins.

To explain this, let us leave the genetic code
for a moment and go back to the biochemistry of
protein synthesis. We already know that proteins
are made on ribosomes. Each ribosome consists of
two subunits, a smaller and a larger one. Each

subunit has one rRNA molecule and 30 to 60 different protein molecules. In a cell free ribosomes
separate into two parts. The "starting" tRNA, with

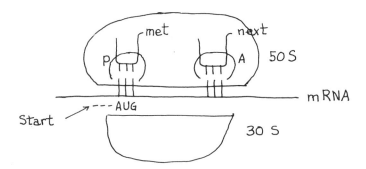

its methionine or formyl methionine, associates
with the small subunit and with the start sequence
of mRNA. Then a large ribosomal subunit joins
in and the complex is ready to go (provided a number of other ingredients are also present; let us
forget about them).

The ribosome has two adjacent sites, A and P,
each of which accommodates a tRNA. Usually a tRNA
with its amino acid would make contact with its
codon at the A site. The "start tRNA" (met or
formyl met) is unique because its structure somehow makes it go to the P site of a messenger ribosome complex, where the "start AUG" codon is waiting. In this way, the starting amino acid becomes
located at the P site, ready to accept the next
amino acid to the A site, according to which codon

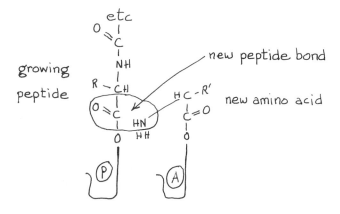

is there. The next step is the enzymatic joining
of amino acid number two to number one. The α-
amino group of the second amino acid attacks the

bond between the first amino acid and the terminal
nucleotide of the tRNA. A peptide bond is formed,
the first amino acid jumps piggyback onto the
second. The polypeptide chain has been started.

Next, the unloaded tRNA on P is shoved off and
the loaded one on A moves to P (or rather, the
ribosome moves one triplet downstream until its A
site is opposite the next triplet). The process
repeats itself till it meets a termination signal.

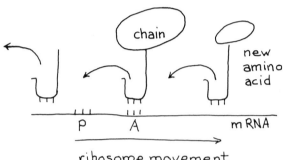

Of course this very complicated machinery requires enzymes and energy. Some of the enzymes
are fixed on the ribosome, others are in the cell
sap. The energy is provided, partly by the high-
group-donor potential of the amino acyl-tRNA bond,
partly by GTP; two molecules of GTP are expended
for each amino acid added. How the energy is used
is still unclear: for example, how does the ribosome move? The whole system resembles a production line: there are still many things to unravel
about its moving drives and control switches.

Polysomes

In a production line, a new car does not wait
to start until the preceding one is finished.
Likewise, as a ribosome slides along a messenger
RNA, the start signal becomes again available for
another start. Under full steam, there can be a
ribosome at work every 200 or 300 nucleotides,
each making a polypeptide chain. A messenger
with its working ribosomes is called a polysome
(see page 108).

An interesting question: does a polypeptide
chain begin to form its three-dimensional structure while it is being synthesized or only after
it is released? This is important, because the
final three-dimensional structure, which after
all is the only thing that matters in a protein,
may differ depending on whether the folding occurs
before the chain is finished or only after it is

completed. If folding occurs continuously, seek-
ing arrangements of minimum free energy for the
early portions of the polypeptide, the structure
thus established may be different from the state
of lowest free energy for the completed polypeptide. For example, a sequence of hydrophobic,
uncharged amino acids located in the late portion
might tend to be hidden in the center of the protein molecule, but if the earlier part of the
chain was already balled up in a tight three-
dimensional structure it would be unable to do so.
Some small protein molecules, after being completely denatured into loose chains, return spontaneously to the original form. But some of the
larger ones may not do so. Because its program
is expressed in a time sequence, the protein molecule made in a cell may not be in the configuration
that a chemist would call the energetically most
sensible form.

Regulation

One more question: are all proteins of a cell
made in equivalent amounts? Certainly not. An
automobile factory capable of making 17 different
car models would be in trouble if it had no way
of modulating how many of each model it made, according to demand, cost, and so on. Likewise, a
cell whose genes worked all at a constant rate
would make too much of some proteins and not enough
of some others. A program is useful only if it
has both instructions of what to make and instructions of how much to make (that is, if it knows
how not to make).

For example, the DNA of E. coli is sufficient
to code for about 4,000 proteins. If all of them
were made all the time in maximum amounts, the
rate of protein synthesis would be 4,000 times
the rate for any one protein. This is definitely
not so. Likewise, a human liver cell or a muscle
cell have the same genes, but the liver cells make
mostly liver enzymes, and the muscle cells make
mostly contractile proteins. Evidently there must
be regulation of gene function. Genes are shut off
and turned on when necessary, either in response
to the external food supply or in response to
developmental stimuli during the growth of a complex organism.

At this point it would be useful to understand
how gene regulation works--one of the most exciting
fields of molecular biology today. But we must
make a detour. We cannot go into the regulation

of gene action before we learn more about genes
and genetics. The reasons will become obvious if
you stop to consider that until now you have
learned general properties of all genes--DNA structure, messenger production, protein synthesis, the
genetic code, the signals to start and end a protein chain. But the individual genes are all different, in length, in sequence, and in regulation.
To understand the experiments that have clarified
these properties we must first delve into the gene
concept and the organization of genetic materials.

Part III

GENETICS

Lecture 14

Gene concept

There are two ways to think of a gene. The first is to think of the sequence of DNA that programs, through mRNA, the sequence of amino acids in a given protein. This concept was embodied in the rule "one gene, one enzyme" enunciated by Beadle and Tatum in 1941 and later more correctly stated, "one gene, one polypeptide chain" (since an enzyme protein may consist of more than one type of polypeptide chain). This biochemical definition of the gene reflected the realization, based on experiments with microorganisms, that gene mutations altered the function of individual enzymes by actually modifying the structure of the enzyme proteins.

The older, classical approach to the gene is one that ignores primary functions and defines a gene as a hereditary factor that affects a recognizable trait of an organism. This was the approach used by Mendel in his famous study of inheritance in garden peas.

Mendel succeeded so well in his studies of inheritance because he thought simply and clearly (not only was he a monk but he was also not teaching at a university). He concentrated on single, discrete differences in well-defined characters such as green seeds versus yellow or white flowers versus red. Previously people had tried to study the heredity of complicated traits such as body height or skin color with little success. Also, Mendel was careful to start with lines of garden peas that had remained constant for several generations. Most important of all, he kept precise quantitative counts of the plants he studied.

Genes

The clear-cut differences in recognizable traits turned out to be explainable by the assortment of stable factors, which had all-or-none effects and were not "diluted" at each generation. Even if they failed to show up in one generation, they would reappear unchanged in the next generation. These factors are what we now call genes.

The organisms appropriate for this type of genetic experimentation are organisms which, for better or for worse, have yielded to the temptation of sex. Each new individual is the result not of splitting off a part of the old individual (vegetative reproduction, as in bacteria) but is the product of the coming together of two germ cells or gametes. Therefore each cell except the germ cells contains two sets of determinants for

each Mendelian trait, one from each parent. In
plants the male and female germ cells are called
<u>pollen grains</u> and <u>ova</u>, respectively: in animals
the corresponding cells are called <u>sperm</u> and <u>egg</u>.

Consider, for example, a human trait controlled
by a single gene, such as the ability to curl one's
tongue (about 85 percent of the population can do
it). Every cell of a person's body contains two
copies of this gene. If you can curl your tongue
it means that you have at least one copy of the
gene in question, which must control the innervation of the transverse muscle of the tongue. (Those
of you who cannot curl might as well stop trying.
Curling cannot be learned if you do not have the
genes.)

Mendel's laws

In his experiments Mendel crossed two "parental
lines" (the P generation) of peas and looked at
their offspring (the F_1 generation). The F_1 offspring were then mated among themselves and the
members of the next generation (the F_2 generation)
were examined. A cross consisted in depositing
the pollen of one plant on the female flowers of
another plant. The parental plants had been grown

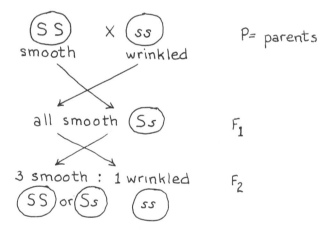

and self-fertilized for many generations so that
their progeny would be uniform.

Mendel's First Law

When plants with smooth seeds were crossed
with plants with wrinkled peas, for example, all
F_1 plants produced only smooth peas. The plants
in the F_2 generation, however, showed both characteristics in the ratio of three smooth to one
wrinkled-seed plant.

This result illustrates what is now called
<u>Mendel's First Law</u>, the law of independent segre-
gation.

Dominance

Mendel explained the results by postulating that the F_1 plants received one gene for smooth (S) seeds from the pure smooth parent (SS) and one for wrinkled seeds (W or s) from the pure wrinkled parent (WW or ss). They are <u>heterozygotes</u>, that is, they contain two different versions of the same gene. The S gene is <u>dominant</u>: it masks the presence of s when both S and s are present; therefore the heterozygotes have smooth seeds. The s gene is referred to as <u>recessive</u>. When the heterozygous F_1 plants produce pollen and eggs, these have one gene copy only: one-half of them has the gene for smooth peas, one-half the gene for wrinkled peas. The outcome of a cross between two F_1 plants is best seen using a checkerboard diagram.

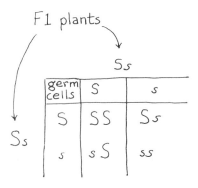

Since S is dominant, the SS and Ss plants produce only smooth seeds and only ss plants produce wrinkled seeds. The checkerboard explains the ratio of three smooth to one wrinkled-seed plant in the F_2 generation as was observed by Mendel.

How can we distinguish SS from Ss, if S is dominant? A <u>test cross</u> (crossing the unknown with a bona fide ss) gives the answer:

$$Ss \times ss \text{ (tester)} \longrightarrow 1/2\ Ss,\ 1/2\ ss$$

$$ss \times ss\ (\ ``\ ) \longrightarrow \text{all ss}$$

Mendel's Second Law

Mendel also studied simultaneously the inheritance of pairs of traits and found that any two traits behaved independently. They were inherited as if each one ignored the other. A checkerboard shows that there are 4 combinations of traits, which appear in ratios of 9:3:3:1. This is

Mendel's Second Law, the law of independent assortment. [This is only a first-order approximation

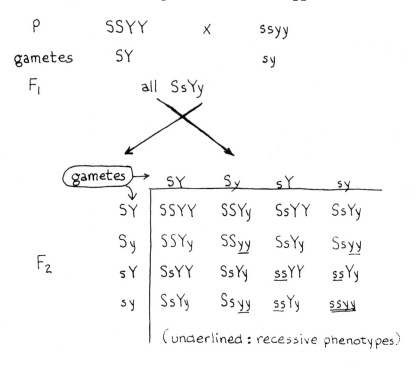

P   SSYY   X   ssyy

gametes   SY   sy

F₁   all SsYy

F₂

gametes →	SY	Sy	sY	sy
SY	SSYY	SSYy	SsYY	SsYy
Sy	SSYy	SSyy	SsYy	Ssyy
sY	SsYY	SsYy	ssYY	ssYy
sy	SsYy	Ssyy	ssYy	ssyy

(underlined : recessive phenotypes.)

rule, and Mendel either was lucky or possibly
cheated a bit by not reporting experiments that
did not fit his theory].

There are some complications, however. Sometimes the F_1 progeny do not look like either of
the two parents. This is because of lack or in-
completeness of dominance. For example, in dealing with flower color, a red-flowered plant crossed
with a white-flowered plant gives a pink-flowered
F_1 generation. The F_2 plants are 1 red; 2 pink;
1 white. The reason is that the red color is due
to the action of an enzyme, and one copy of the
red gene per cell does not make enough enzyme to
produce a full red color, but it does make enough
for a pink color. All sorts of similar cases
arise since the final level of expression of a
given trait depends on how many copies of the corresponding enzyme are made, how efficient they are,
etc. Only when one gene copy is as efficient as
two is dominance complete.

Two genes acting on the same developmental process may act more or less independently. For example, what makes a pea wrinkle? The pea is dis-
tended by starch accumulated in it: if there is

either less starch or a different kind of starch, the pea is wrinkled. The color of the pea seed, on the other hand, is due to several enzymes, one of which converts the outer seed skin from yellow to green. If the enzyme is missing, then the pea is yellow. Since these two characters are controlled by different enzymes in separate pathways they behave independently. But this is not always the case: often the action of one gene affects the expression of other genes. It is this kind of thing that renders genetics difficult and wonderful at the same time.

Diploid versus haploid

Organisms like peas or corn or _Drosophila_ or man are <u>diploid</u>: most cells have two copies of each gene. Organisms like bacteria are <u>haploid</u>; their genetic material consists of only one copy of each gene. Other organisms, including algae and fungi (molds and mushrooms), are mainly haploid but occasionally spend part of their life in the diploid state.

Why did evolution produce so much variety of species among those organisms that are diploid for most of their lives? One advantage of diploidy is that the mating process causes the set of genes from one parent to mix with the set from the other parent. In this way new and possibly more advantageous combinations can occur and be tested by natural selection.

Heterozygosis

Another reason why it may be advantageous for an organism to spend a large portion of its life in the diploid state is that heterozygosity _per se_ seems to be advantageous. That is, it is often better to have two different copies of a given gene than to have two identical ones (This advantage is called <u>heterosis</u> or hybrid vigor.) Heterozygosity also protects mutant genes that arise by mutation and allows them to be tested in a variety of different combinations, as genes are reshuffled at each generation.

There are many crosses of the two-trait type that do not give 9:3:3:1 ratio reported by Mendel in his crosses of garden peas. For example, two traits may hardly segregate at all so that the results (3:1 in the F_2 generation) are like those expected if both traits were due to the same factor. Or, the four types predicted by Mendel may be present in the F_2 generation, but in frequencies different from 9:3:3:1. Many of these deviations were discovered when Morgan and his colleagues

started working on the genetics of the fruit fly
Drosophila melanogaster, a much easier organism
than peas: it lives in milk bottles with corn
meal as food and two generations are produced in
only two weeks. (Not quite as fast as bacteria,
but much more sex going on.)

The great advantage of _Drosophila_ was, of
course, that given the large numbers of flies
available one could find mutants and do crosses between mutant lines derived from the same stock
rather than between plants whose differences
stemmed from unknown events in the field. The
frequency of new mutants can be increased by x
rays, as discovered by H. J. Muller in 1927.

Linkage

Morgan already knew one more thing about genes:
they were located in the chromosomes of the cell
nucleus. Soon the deviations from the 9:3:3:1
rule were interpreted as follows. If two traits
assort independently, giving 9:3:3:1 ratios, their
genes are usually in two different chromosomes.
If two traits do not assort independently, they
are said to be _linked_: their genes are presumed
to be located in the same chromosome. The linked
genes occasionally become separated by a mechanism
different from the random assortment of chromosomes: that is, by exchanges of parts between
homologous chromosomes. The closer together two

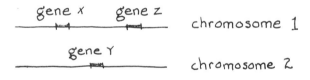

1) X and Y or Z and Y assort independently

2) X and Z are linked (non-random assortment;
recombination requires
crossing-over).

genes are, the more infrequently they become
separated.

Note that even without invoking chromosomes
one could construct a purely formal model of
linkage. Inherited traits are determined by
genetic loci; if two traits are assorted independently, their loci are unlinked; if they do not,
they are linked. The mechanical, chromosomal interpretation adds one important feature: if a
chromosome is "long," that is, has very many sites

where exchanges can occur, genes located very far on the same chromosome behave as if they were un-linked. (Do you see why? When many exchanges occur, the chance of having 2n or 2n + 1 exchanges--even or odd numbers--is the same; hence the distant genes separate in 50 percent of the cases on the average, just as genes in different chromo-somes do.)

Further reading   Strickberger's Chapters 6 and 7, "Mendelian Principles, I" and "II," and Chapter 16, "Linkage and Recombination," are a more complete treatment of formal diploid genetics.

Lecture 15

Crossing-over

While geneticists were doing crosses with peas, corn, mice, and fruit flies and identifying genetic loci, the cytologists were studying chromosomes and their behavior in dividing cells and in the production of germ cells. They saw that in the latter process, called _meiosis_, there were exchanges of parts between two chromosomes that looked more or less alike and could be the pairs of _homologues_ derived one from the father, the other from the mother.

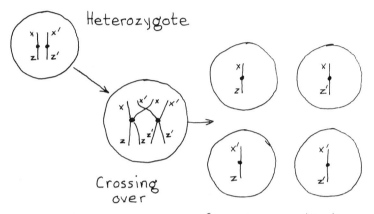

four germ cells, two parental, two recombinant

It was only a small jump to postulate a connection between these exchanges and gene linkage. If two parents containing genes xz and x'z', respectively, are crossed, and if genes x (or x') and z (or z') are in the same chromosome, xz (or x'z') remain together in the offspring unless there is an exchange, or _crossing-over_, between the two genetic loci. The closer together the genes are on the chromosome, the less likely they are to separate by crossovers between them--_if the incidence of crossing-over is more or less proportional to linear distance on the chromosome._

This model of crossing-over was verified experimentally by Barbara McClintock using corn and by Curt Stern with fruit flies. [Before reading ahead, take paper and pencil and try to figure out what is to be proved and what constitutes a proof.] McClintock constructed genetically marked strains which were also cytologically marked. That is, she could distinguish under the microscope the two homologous chromosomes of corn suspected of carrying the pair of genes under study--for example, x

and z versus x' and z'. The cytological markers
were knobs stuck on the chromosomes because of
previous x-ray treatments.

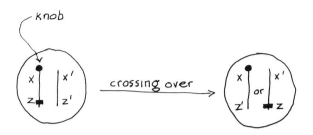

Recombination

Crossing-over between the genes should not only
create germ cells xz' or zx', but also create new
configurations of chromosomes in those descendants
in which genetic <u>recombination</u> between the x and z
genes had occurred. The results were entirely in
accord with the model: crossing-over involves
exchange of parts between two chromosomes (or,
rather, between two <u>chromatids</u>, that is, the
chromosomal threads that will go to the germ cells).

Mitosis

At this point we must look a bit closer at the
process of <u>mitosis</u>--cell division in eucaryotic
cells--and its relative, meiosis--cell divisions
leading to production of germ cells--from the
chromosome point of view.

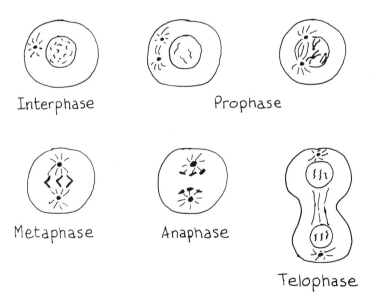

Mitosis includes several steps:
1. <u>Prophase</u>. The chromosomes, which are generally

invisible in <u>interphase</u>, become visible as fine threads. Each chromosome is already split into two chromatids, which are the result of DNA replication and chromosomal duplication. The chromatids are joined together at one point called the <u>centromere</u>; the nuclear membrane disappears, and a spindle appears consisting of protein microtubules.

2. <u>Metaphase</u>. The split chromosomes condense and line up in the middle of the cell, with centromeres toward the center.

3. <u>Anaphase</u>. The chromosomes pull apart, one copy of each chromosome going toward each pole, migrating along the spindle toward the <u>centrosomes</u>.

4. <u>Telophase</u>. Migration is complete, the nuclear membrane reforms, the cytoplasm is split, and the chromosomes fade out. Interphase starts again.

Thus after each mitosis each chromosome has become two, and one copy has been distributed to each of two daughter cells, which therefore have a full genetic complement as the parent cell had.

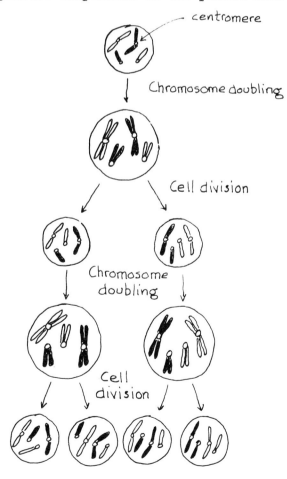

Mitosis

There has been no change in gene sequence, no segregation, no recombination.

What are the biochemical correlates of mitosis? In actively reproducing cells--mitosis every 24 hours, for example--the actual process of mitosis (M) may take about two hours.

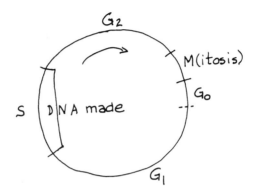

Then for about 6 hours (G_1 period) the cell grows, proteins, ribosomes, etc. are made, but nothing seems to happen to the chromosomes (except of course, that their DNA serves as templates for RNA synthesis). Next comes the S period, about 7 hours, in which new DNA is synthesized and the amount in the chromosomes doubles. This is easily seen by following the incorporation of radioactive thymine by autoradiography.

Then DNA synthesis ceases but mitosis does not yet start. There is a period (G_2) when the chromosomes stay put but already have double DNA: that is, there are two copies of each gene in each chromosome. Mitosis, therefore, is just the separation of the DNA copies already made during the S period.

Note that in this whole process there is only one critical trigger, just at the beginning of G_1. Cells such as brain or muscle cells that are not dividing are always stopped at the beginning of G_1, a point also called G_0. If a stimulus gets them started, they go through the whole operation and stop again at G_0. They never stop, say, just before mitosis.

Now let us come back to Mendel's rules. If body cells have two copies of each gene, that is, two copies of each chromosome (diploid) whereas germ cells contain only one copy (haploid), how does this <u>reduction</u> or <u>meiosis</u> take place? There must be a mechanism for systematically reducing

the number of chromosomes. The process of meiosis
has the same general phases as mitosis but is
profoundly different.

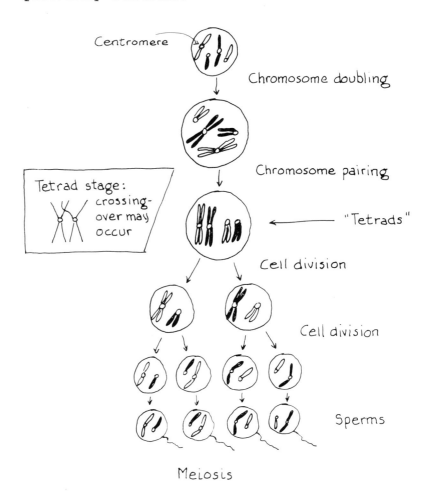

Meiosis

1. The first steps of meiosis are the same as
in mitosis: the chromosomes of the cells that
will generate the germ cells (for example, <u>sperm-
atogonia</u> preparing to produce four sperms) become
visible, the nuclear membrane disappears, and
each chromosome separates into two chromatids
joined together at the centromere. As in mitosis,
DNA replication has already occurred. There is
no further DNA synthesis during the meiosis process.

2. Next, each doubled chromosome <u>pairs with
its homologue</u> forming a <u>tetrad</u>: two paired chromosomes, each split into two chromatids but with un-
split centromeres. At this stage, corresponding
to metaphase, crossovers take place: visible ex-

changes of parts, which can separate genes located in the same chromosome.

3. In the next stage one centromere of each tetrad moves to each end, taking with it its chromatids (which may have undergone exchanges). The cell then divides (<u>I meiotic division</u>).

4. Next, the centromeres split, the two chromatids move to opposite poles, and the cell divides again (<u>II meiotic division</u>).

In the overall process each tetrad generates four haploid cells, each receiving one chromatid. These haploid cells, (all of them or only some, depending on species and on sex) become the germ cells, which fuse when they meet and generate a diploid <u>zygote</u>. The zygote is the starting cell of the new organism.

Notice that the formation of haploid germ cells explains the reappearance of recessive characters in Mendel's First Law: an Aa heterozygote generates both A germ cells and a germ cells. Mendel's Second Law reflects the fact that at the I meiotic division the two chromosomes from a tetrad move to one or the other pole at random, so that a parent Aa Bb (A and B on <u>different chromosomes</u>) can generate germ cells AB, Ab, aB, or ab with equal frequencies.

Crossing-over, in addition, can scramble genes in the same chromosome. AC in one chromosome and ac in the other one can produce four combinations, the proportion of new ones (Ac and aC) depending on how far the A and C genes are.

The sum of the two processes--random reshuffling of chromosomes plus crossing-over--gives rise to <u>genetic recombination</u>, that is, new combinations of genes in the next generation. It is this reshuffling of the genome at each generation that makes evolution effective by making available innumerable combinations of genes for selection to act upon, and especially by permitting any new gene arising by mutation to be tried in a variety of genetic settings.

In diploid organisms, mutations that occur in somatic cells are not easily seen. Those that occur in the germ line may remain masked generation after generation if they are recessive. There are exceptions, for example, for mutations in the sex chromosomes--but we shall leave sex alone for the time being. To study the mutation process, and especially to use mutations and the

recombination of mutated genes to unravel the
secrets of gene action, it is convenient to concentrate first on organisms that spend most or
all their life in the haploid state. So we leave
peas, fruit flies, and Homo sapiens to concentrate
for a while on two simpler organisms: the bread
mold Neurospora crassa and our old friend E. coli.

Neurospora grows well in a simple, minimal
medium of salts, sugar, plus one vitamin (pantothenic acid, a component of coenzyme A: remember?).
Like E. coli in a similar medium, the mold must
synthesize everything else it needs. The mold
nuclei are haploid. A gene mutation that blocks,
for example, a step in the synthesis of an amino
acid is lethal in that medium: but if you provide
that amino acid everything is fine. Compare that
with our own inability to synthesize 12 amino
acids and 6 or more vitamins: they must be
present in our food, or we perish.

Why Neurospora provided the first step toward
a nutritional genetics will become clear in the
next lecture.

Further reading   Strickberger's Chapter 2, "Cellular Division
and Chromosomes," should be a useful supplement.

Lecture 16

Life cycles

In higher plants and in most animals the whole
organism consists of diploid cells. Diploidy
arises as a result of the fusion of two haploid
cells, the sperm and egg (or pollen nucleus and
ovum in plants). Throughout the life of the organism diploid cells reproduce by mitosis, forming new diploid cells. Only in the sex organs
the organism produces haploid germ cells by meiosis. We can describe the <u>generalized life cycle</u>
of organisms with a diagram:

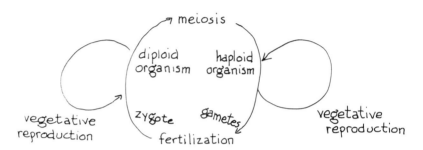

The left-hand loop represents the growth and development of the diploid section of the organism; the
right-hand loop represents the haploid germ cells
or gametes. [In most animals the ancestors of the
germ cells are set aside early in development,
while in plants any part of the adult organism can
give rise to germ cells.] When gametes are fertilized by other gametes the diploid cycle begins
again.

In both higher plants and animals, the right-
hand loop is reduced to the germ cells only: the
haploid cells do not multiply as such (except once
in the pollen, as you shall learn later). Other
organisms, however, have a more extensive haploid
stage: the haploid cells divide and constitute a
large part of the organism. The life cycle includes extended reproduction in the haploid stage.
All possible combinations exist in nature, from
those in which the diploid loop is predominant to
those in which most of the life cycle occurs in
the haploid state.

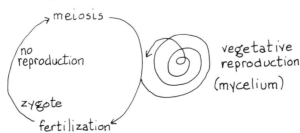

The latter is true of Neurospora. Like most
fungi, this organism consists mostly of a mass of
filaments called a mycelium, with haploid spores
called conidia. There are two "sexes," A and a.

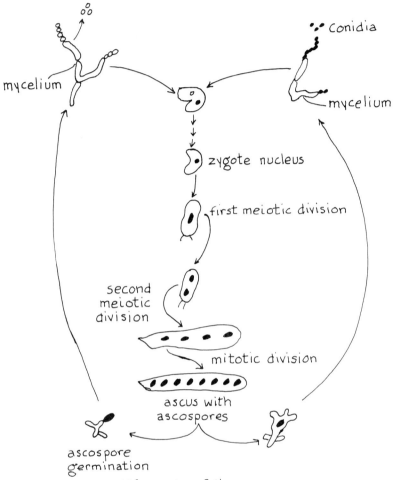

Life cycle of Neurospora

If A meets a, filaments or conidia fuse, then two
nuclei fuse to produce a diploid nucleus or zygote,
which goes immediately into meiosis. So in this
organism practically the whole life is lived in
the haploid state: the diploid stage lasts only
one cell generation.

It is because of haploidy that Neurospora is so
useful for isolating and characterizing nutritional mutants, that is, mutants that require some
specific nutrient to grow: they grow on rich media
but not on media without the required food sub-

stance. These mutants can be used to deduce biochemical pathways by feeding to them a variety of
suspected precursors of the required nutrient.
And they can be used for crosses to construct a
genetic map locating the genes for specific
enzymes on various chromosomes. The frequency of
mutants can be increased by exposing the organism
to a mutagenic agent. Ultraviolet light, x rays,
and various chemicals are mutagenic because they
cause chemical changes in the DNA. When the
altered DNA undergoes replication, the incidence
of errors of replication increases, and mutants
are produced.

Nutritional mutants   In 1941 Beadle and Tatum started to develop
the genetic system of Neurospora to study biochem-
ical genetics. They proceeded to reconstruct biosynthetic pathways by isolating, for example,
several mutants that required a given amino acid
X, then establishing which of them could grow also
with presumed precursors A,B,C,...

$$A \rightarrow B \rightarrow C \rightarrow \ldots \rightarrow X$$
$$ \quad 1 \quad\; 2 \quad\; 3$$

Mutants blocked at step 2, for example, may grow
with either X or C, provided C can enter the cells.
Each step turned out to be controlled by a gene or
by two or three genes if the enzyme consisted of
two or three different polypeptide chains.

I already mentioned that Neurospora spends most
of its life cycle in the haploid phase. When
mating does take place between conidia or filaments
of opposite mating types, A x a, meiosis occurs
immediately and produces eight ascospores contained
in an ascus:

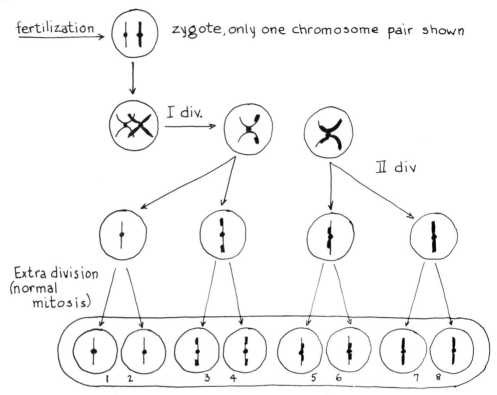

Spores in an ascus, in identical pairs

(There are eight spores rather than the usual four products of meiosis because there is an extra division; but this has no genetic significance.) One can pull the spores out of the ascus in serial order and observe all products of meiosis. Four of the spores would be of mating type A, the other four, a: the mating type is due to one gene only. The mold growing from each of the spores is then tested, for example, for its ability to grow in a nutrient mechanism lacking in food substance that is being tested, such as an amino acid X or one of its precursors.

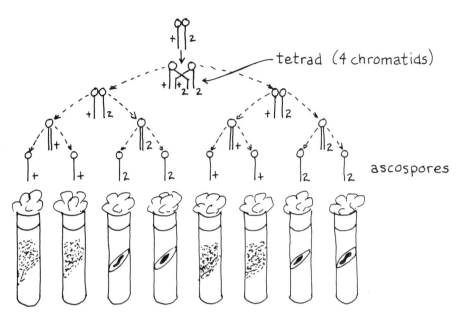

tetrad (4 chromatids)

ascospores

Mapping function   For genes in different chromosomes, genetic
differences or <u>markers</u> assort independently, as we
already know. Two markers on the same chromosome
are inherited together unless they have become
separated by crossing-over during meiosis. Thus
the 8 molds that grow from the 8 spores of an
ascus derived from a cross between two <u>Neurospora</u>
lines that differ in several characters can show
a large amount of recombination. Suppose in a
cross XYZ x xyz the three genes, X,Y,Z, distribute
themselves at random, so that by studying many
asci you find 50 percent parental descendants and
50 percent recombinants for each pairwise combination. You conclude that X,Y, and Z are in different chromosomes. If, instead, you find that X
recombines with Y 10 percent of the time and Y
with Z 5 percent of the time, you conclude that
they are all linked, that is, in the same chromosome. If X recombines with Z 13 percent of the
time, you can begin to draw a map as follows:

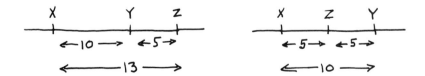

If X-Z recombination is 5 or 6 percent, you draw
a different map. Recombination frequencies are

additive, but only approximately, because there
can be more than one crossover occurring in the
same intervals. Two crossovers, one between X
and Y and one between Y and Z, restore the parental combination of X and Z and cause an apparent
decrease in recombination. The following equation

$$\overline{XZ} = \overline{XY} + \overline{YZ} - 2(\overline{XY} \times \overline{YZ})$$

where $\overline{XY}$, etc., are the frequencies of separation
between pairs of genes that were together in the
parent, is called a mapping function. [Note what
happens in a three-point cross XYZ by xyz: the
least frequent classes are XyZ and xYz because
they require two crossing-overs.]

Linear maps
When all data obtained not only in Neurospora,
but in any organism, from viruses to man, are
fitted to this equation, it turns out that all
linked genes can be placed on topologically unbranched maps. Mutant forms of a gene are
called genetic markers. The order of genetic
markers is established by the recombination frequencies. Note that in this analysis one makes
the assumption that crossovers are symmetrical
and reciprocal, that is, that the pieces that are
exchanged are of equal length and carry the same
gene loci. The experiments with single asci of
Neurospora confirm this: all the expected combinations are present, with no loss of one type or
another.

Genetics of E. coli
There are shortcomings to working with Neurospora, however. First, it shares with all fungi
the quality of having a tough skin: it is hard
to get out the enzymes, so that the biochemistry
is not well known. Also, even though Neurospora
grows relatively fast and has excellent sex, it
cannot be handled in very big numbers. Therefore,
around 1943 people started doing genetics with
bacteria. This work, especially with E. coli, went
ahead at such a pace that by 1965 it became a
feasible goal to identify in biochemical terms
every single gene of E. coli (the so-called "Project K"). By now, in 1973, almost 1,000 known
genes of E. coli have been located on a chromosomal map.

Isolation of mutants
Mutants of E. coli are readily isolated on
selective media: plates of agar medium on which
the normal cells do not form colonies but certain
mutants do. For example, if you incorporate in
the agar 0.1 mg/ml of the antibiotic streptomycin,

you select out streptomycin-resistant mutants
(about 1 for 10^9 cells plated). Or, you can use a
bacteriophage, that is, a virus that kills all
cells except a few resistant mutants. And so on.

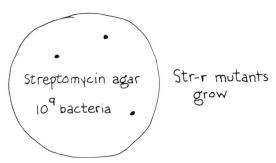

Streptomycin agar
10^9 bacteria

Str-r mutants
grow

You can also isolate mutants that <u>cannot</u> grow
in a certain medium. For example, you want a
mutant that cannot synthesize the amino acid tryptophan. You put your bacteria in a medium with
sugar, salts, a mixture of amino acids, but no
tryptophan. Then you add <u>penicillin</u>. Penicillin
is a strange antibiotic: it blocks the synthesis

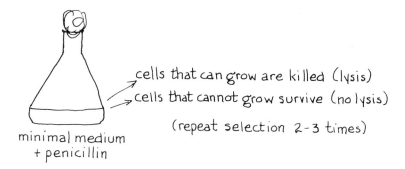

cells that can grow are killed (lysis)
cells that cannot grow survive (no lysis)

(repeat selection 2-3 times)

minimal medium
+ penicillin

of the cell wall and, therefore, it kills only
the cells that are growing because their cytoplasm
bursts out of their skin. Those cells that cannot
grow remain alive, and these include the cells
that cannot make tryptophan. In this way you can
enrich your culture of tryptophan-negative (<u>trp</u>⁻)
mutants and, in a few repeat experiments, isolate
them.
     With thousands of mutations available in <u>E.
coli</u>, even apart from sex, it should be possible
to learn quite a bit about genetic control of
enzymes and pathways. You can isolate <u>trp</u>⁻
mutants that behave as mutants at 40°C but as
normal cells at 25°C. Then you find that one of

the enzymes in the tryptophan pathway in the
mutant is temperature sensitive (ts):

$$A \rightarrow B \rightarrow C \overset{ts}{\rightarrow} D \rightarrow Trp$$
$$\quad 1 \quad\; 2 \quad\; 3 \quad\quad 4$$

Or, you may find that a mutant blocked in step 4
excretes substance D in the medium. This can help
identify D as a precursor of tryptophan. If the
mutant that excretes substance D is laid down
side-by-side with a mutant blocked in step 2, substance D may help this second mutant to grow
(cross-feeding or syntrophism).

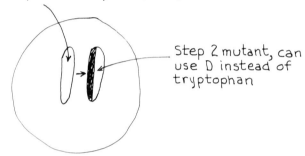

Syntrophism

All these tricks and many others are the routine
tools of bacterial genetics. An important point:
by repeating the isolation of mutants one after
the other with the same bacteria one can prepare
strains with 2,3,..., n mutations for a variety of
experiments--including sex. For example, one can
construct a line of E. coli called K-12 trpA⁻
isoE^{ts} str^{r} ton^{r}--meaning a derivative of the K-12
strain of E. coli that requires tryptophan (be-
cause of a mutation in enzyme A of the trp pathway); that requires isoleucine at high temperature
(because of a ts mutation in isoleucine enzyme E);
that is resistant to streptomycin; and that is
resistant to phage T1.
     Some mutations are lethal because they affect
essential genes: for example, they may block DNA
synthesis, or protein synthesis, or cell wall production. But even such mutations may be temperature dependent, and then the mutants can be saved
and studied by keeping them at the temperature
where they can grow and testing samples of their

cultures at the other temperature. This is the
way to study, for example, the genes that control
RNA polymerase or the enzymes of protein synthesis.

Are bacteria haploid or diploid? Mutation
studies by themselves suffice to prove that they
are haploid, because mutations produced by mutagenic agents such as ultraviolet light, x rays,
or certain chemicals are expressed immediately
and in both directions. If bacteria were diploid,
there would be two gene copies. For trp^- to express itself it would have to be dominant; but
then the reverse mutation, from trp^- to trp^+ when
observed in a "diploid" trp^- strain could not express itself.

The bacterial life cycle does not fit the
generalized life cycle scheme we have used. There
is no cycle, in fact, since there is no regular
production of gametes nor alternation of haploid
and diploid phases. Most bacteria apparently live
forever without mating, that is, without exchanges
of genetic material. You can easily see that this
limits the avenues for evolution: new forms can
arise only by mutations--either changes in the
genes or losses or duplications of genes--without
the chances for reshuffling of genetic material
upon mating.

Some bacteria, however, have found ways to
escape from the doldrums of sexless life. At
least, they are capable of exchanging genetic
material. This may or may not have played much
of a role in bacterial evolution, but it has certainly played a big role in the analysis of gene
action and the organization of genes in the DNA.
[We have no idea yet how much transfer of genetic
material there is among becteria in nature, and
we shall let it go at that.]

**Transformation by
DNA**

The first kind of genetic transfer to be discovered in bacteria was a wholly artificial one
and was also the most momentous. As we already
mentioned in connection with the molecular structure of DNA, Avery, MacLeod, and McCarty discovered
in 1944 that in the pneumonia bacilli they could
produce transformation of genetic characteristics
by means of DNA.

pneumococcus R   + DNA from pneumo. S → 0.1 — 10% pneumo S

This was the first direct evidence that DNA was the genetic substance of bacteria or, for that matter, of any organism. Transformation can be observed and studied in certain species of bacteria besides the pneumonia bacilli, but not in E. coli, because the latter normally cannot take up free DNA. Transformation has been studied in detail especially in Bacillus subtilis. What happens is that the DNA enters, and a piece of DNA, generally one strand alone, replaces the corresponding "resident" strand in the DNA. Genetically speaking, if the recipient bacterium was, for example, met$^-$thr$^-$ and the donor bacterium, the one that provides the DNA, was met$^+$ thr$^+$ some bacteria become transformed to thr$^+$met$^-$, some to thr$^-$met$^+$. Since only one strand enters, at first the cell is heterozygous, for example, met$^+$/met$^-$; but after one DNA duplication it gives one met$^+$ and one met$^-$ cell. The reason is that only small pieces of DNA enter the recipient cell and usually carry only one genetic marker at a time. Since only a few of the bacteria receive any pieces of donor DNA that carry a particular marker, the chances of two genetic markers from the donor accidentally entering the same recipient cell are relatively low.

Occasionally, however, two genes of the donor are introduced jointly on the same DNA piece, because they are linked together. The frequency of this joint transformation is a measure of how close two genes are to each other on the DNA of the bacterium. In this way one can construct a fully consistent genetic map.

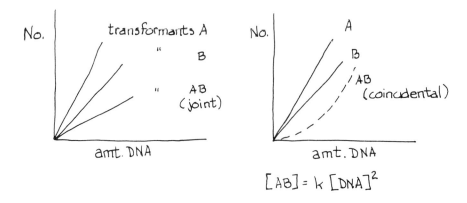

$$[AB] = k\,[DNA]^2$$

(To decide whether two genes of the donor found together in the recipient have entered together or coincidentally, you compare the joint-transformation frequency with the single transformation frequencies.)

It turns out that <u>all known genes of B. subtilis can be fitted on a single chromosome</u>, and <u>that chromosome is circular</u>. That is, if the genetic markers are called A.B.C...X.Y.Z, then A is occasionally transferred with B on the same DNA fragment; B is transferred with C; and so on; and Z is transferred with A! The same situation turns out to hold for all other bacteria: a single circular chromosome with all genes upon it in agreement with Cairns' autoradiographic finding of DNA

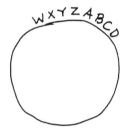

circles in <u>E. coli</u> extracts!

Further reading

Strickberger's Chapter 3, "Reproductive Cycles," expands on the subject of life cycles. Stent's Chapter 7, "Transformation," is interesting for technical and historical information.

Lecture 17

Mating in E. coli

E. coli, which has no good DNA transformation mechanism, has something better: real mating. Lederberg and Tatum found that if they mixed together cultures of two E. coli mutants and put them on media on which neither of them could grow, some few colonies came up. For example, a mixture of a strain ton^r str^S, which is resistant to a phage and sensitive to streptomycin, grown together with a strain ton^S str^r, produced some cells that grew in the presence of both phage and streptomycin. These cells were recombinants and their progeny was fully ton^r str^r. Similar results are obtained if one mixes, for example, thr^- and leu^- mutants: the cells that can grow on minimal media are recombinants thr^+ leu^+. Innumerable tests have confirmed that these recombinants are produced by cell-to-cell mating and transfer of genetic material--DNA--between mated cells.

Mating groups

All strains of E. coli fall into one of three "mating groups"--F^-, F^+, and Hfr--which exhibit characteristic mating behaviors. Mixing two F^- strains yields no recombinants, even though the two strains had suitable genetic markers. Mixing

$F^- \times F^-$ no recombinants
$F^- \times F^+$ (some recombinants)
$F^- \times Hfr$ many recombinants

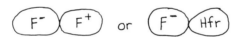

properly marked F^- cells with a culture of F^+ bacteria yields some recombinants. Mating an Hfr (high frequency of recombinants) strain with an F^- produces many recombinants, sometimes almost as many as the number of Hfr cells used in the mixture. An F^+ culture can spontaneously generate Hfr bacteria. Microscopic observations of mixed cultures of F^+ and F^- or Hfr and F^- bacteria show many pairs of cells that remain in contact for 1 to 2 hours and then separate.

What is the difference between these three mating types? The F^- cells contain all the cell's genes in a single chromosome. The F^+ cells contain the F^- chromosome plus an additional small circular piece of DNA--a minichromosome or episome--

called the F factor, which has the genes needed
to make a mating apparatus. The F factor DNA can
be distinguished from the main cellular DNA because it has a different GC/AT content and therefore a different density.

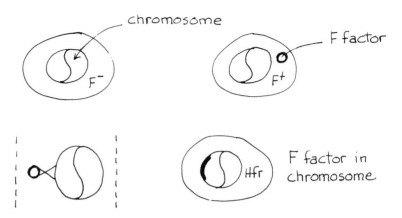

The mating apparatus of the F^+ cells consists
of protein tubes or <u>pili</u> attached to the cell surface. The pili form mating bridges between F^+ and
F^- cells. It is believed, although this is not
quite certain, that the F factor DNA goes through
the hole in the pili from F^+ to F^- cells.

Hfr cells originate when the circular F factor
becomes integrated into the circular chromosome
(without loss of any genetic material). When Hfr
mates with F^-, a piece of the chromosome is transferred that contains part of the F factor DNA and
part of the bacterial cell DNA.

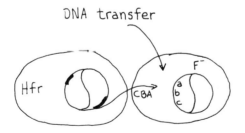

How do we know that DNA and only DNA is transferred in these matings? We can simply label Hfr
or F^+ cells with labels that go either into proteins or DNA or RNA and, after mating these cells
with F^- for 15 or 30 minutes, selectively destroy
the Hfr or F^+ cells and analyze the F^- cells (or
<u>vice versa</u>). We find that nothing passes from F^-

to F⁺ or Hfr; and only DNA passes from Hfr to F⁻ (much less, of course, from F⁺ to F⁻; only the minicircle).

The recipient F⁻ cell is in the same situation as a cell that has received a fragment of free DNA in transformation experiments. It is at first a <u>partial diploid</u> for some part of the genome. If the diploid part is heterozygous for one or more genes, there will be production of recombinants, in which one or more genes from Hfr have replaced those from F⁻. (Of course, the recombinants may have to be selected out on special growth media in order to detect and count them.)

Once you have <u>E. coli</u> recombinants produced by mating, for example, an Hfr with many distinctive genetic markers A, B, C,... with an F⁻ with the corresponding markers a,b,c, you can construct at least a partial map by finding out how frequently A appears together with B, etc.

But the Hfr's, even those derived from the same F⁺, are not all the same: the F factor may have become integrated in any one of many different places in the DNA chromosome--presumably in those places where there is a suitable sequence of nucleotides. This generates a set of Hfr's, each of which transfers to F⁻ cells the same genes but in different orders and even in different directions, as expected from a circle of DNA.

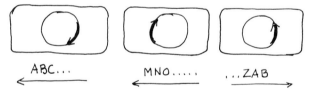

ABC...   MNO.....   ...ZAB

3 Hfr's derived from the same F⁺

How is this complicated story analyzed experimentally? By interrupted mating: that is, mixing the bacteria to be mated, waiting until they have formed pairs and begun transferring DNA, and then at various times (5, 10, 20,...minutes) shaking them apart in some device like a Waring Blender which breaks the mating pairs apart. Then the bacteria are plated on appropriate growth media that select out colonies of presumed recombinants, and these are analyzed for other recombinant characters.

Note that when F^- mates with F^+, the F factor is transferred and becomes established as a free episome. The F^- has become F^+; sex in E. coli is a bit like a venereal disease. The F^+ does not become F^-, it does not lose its F factor, because what is transferred in matings, whether it is the F factor or a chromosomal segment, is always a newly made, single-strand piece of DNA.

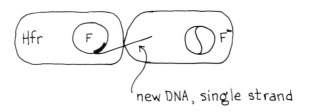

new DNA, single strand

Genetic fine structure

The power of genetic analysis with bacteria is in the enormous numbers of individuals that can be sampled. For example, after a mating involving, say, 10^9 cells of F^- and 10^9 Hfr's, one can detect and count as few as 1 to 10 recombinants. In this way, one can construct maps that reach into what is called, by analogy with spectroscopic analysis, the fine structure of the genetic organization. In fact, one can "atomize" the gene.

Imagine we get a set of E. coli mutants, all of which require, for example, tryptophan, and all of which are defective in the same enzyme--therefore, presumably, in the same gene. Let each of these mutants be present in an Hfr or in F^- and mate them all pairwise. We can then construct a map of the gene based on frequency of recombination. Each mutant corresponds to a mutant site in the map.

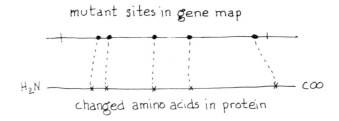

As always, the map is linear. Even more important: if we can, from the series of mutants, isolate the set of proteins corresponding to the mutant genes in question and sequence their amino acids, we

obtain <u>a protein map that is fully colinear with
the genetic map</u>. The farther apart two amino
acids are, the more frequently recombination
occurs between the mutants altered in those amino
acids. This, of course, is simply a corollary of
what we have stated (but not proved) earlier,
that adjacent DNA and mRNA triplets code for adjacent amino acids.

Is there an ultimate limit to the fine structure as defined by recombination? Of course there
must be: there cannot be recombination <u>within</u> a
single nucleotide. But we can postulate that recombination occurs between adjacent nucleotides.

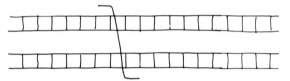

crossover between adjacent nucleotides

Consider now the following mutants, with different amino acids at the same spot in the same
enzyme: Normal (that is, wild-type)--glycine;
mutant I--arginine (enzyme inactive); mutant II--
glutamic acid (enzyme inactive). If one crosses
Hfr mutant I x F⁻ mutant II, or vice versa, one
finds some recombinants with normal enzyme, with
glycine at the proper place. Interpretation:
Mutants I and II had mutations in different nucleo-

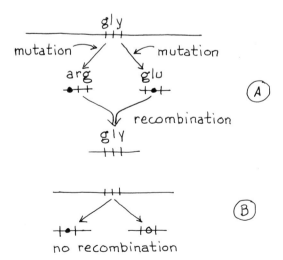

tides of the same triplet; recombination reconstituted the original triplet. The genetic code
tells you this is ok: GGG (glycine) can mutate
to AGG (arginine) or to GAG (glutamic). Then
AGG x GAG can recombine to produce GGG again.
(Remember that recombination occurs at the DNA
level, and that what one observes is the coding
by the messenger RNA produced by the recombinant
DNA genes.)

The recombination between adjacent nucleotides
may occur in only 1 cell out of 10^7 mating pairs
of E. coli. That is the bottom limit. If two
mutants in the same amino acid do not give any recombinants we must assume they carry mutations
that have changed the same nucleotide. For example, GGG can give rise to AGG and to CGG: no recombination is possible.

Partial diploidy

You realize, of course, that the recombinant
bacteria isolated in the experiments just described
are still haploid. They are those in which some
genes from the donor had replaced resident genes
by recombination, the discarded piece being lost.
But there is no end of tricks to bacteria, or
rather, to the ingenuity of bacterial geneticists.
One of these tricks is to produce diploid or rather
partial diploids, by means of which all sort of
problems can be studied.

The key to this trick is the Hfr. Usually, a
given Hfr transfers to the F⁻ cell a piece of DNA
always with the same sequence of "head genes."
Occasionally, however, it may transfer "tail
genes." If one isolates the recombinants that
have received tail genes, one discovers among
them some partial diploids or F' types.

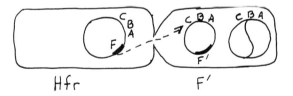

What happens is that the F factor that was in
the Hfr chromosome occasionally comes out carrying
with it a piece of bacterial DNA. Since it is an
F factor, it can remain and replicate independently in the cell and confers upon it the mating competence of F⁺ or Hfr cells. In this way the new
cell line, called F' because it carries an F'

factor, is partially diploid since it has two
copies of certain genes, one copy in the chromosome, one in the F' factor (but in a mating it
transfers those genes that are in the F' factor
rather than the whole gene string). By various
tricks one can construct all sorts of F' factors
with different sets of genes.

Complementation

What good does this do? Consider the case of
two mutants that require the same nutrient, for
example, arginine. Are the mutations in the same
gene or in different genes? If you can construct
a diploid with one copy of each of the mutant
genes, you know right away the answer: mutants
in different genes complement one another: a
diploid A^+/a^-, B^+/b^- has one good copy of each
gene. If the mutations are in the same gene, the
diploid is a^-a^- and there is no function present
(this assumes, of course that a^- and b^- are recessive).

Complementation provides an operational definition of a gene: a discrete coherent chromosomal
region that is responsible for a specific unit
function in the cell. At least one functional
copy must be present. If two mutations affecting

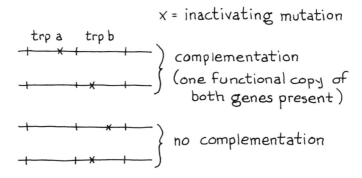

the same function are located in the same gene,
they are said to be cis (from Latin for near); if
they are located in different genes (by complementation tests) they are said to be trans. A
genetic element or gene defined by such tests has
been called a cistron.

Complementation is a very powerful tool for
genetic analysis. It is analogous to a test for
dominance in diploids. It serves to decide
whether sets of mutations that alter the same
phenotypic character are in the same gene. Even
more important, it helps decide which of two forms

of a gene is functional, producing a substance
that can act in trans, and which is the one that
fails to produce a product.

Let us now illustrate the power of complementation analysis by the study of one of the most remarkable set of experiments in molecular biology.

Enzyme induction and repression

These were experiments by Jacob and Monod,
which revealed how the function of individual
genes or groups of genes in bacteria is regulated
in response to the needs of the bacterium. Even
though, unfortunately, we do not yet know to what
extent these regulatory mechanisms apply to cells
other than bacteria, the experiments represent
one of the most elegant feats of biological
analysis.

Begin with a culture of E. coli in meat broth.
In these cells, there are very small amounts--a
few molecules per cell--of a set of three proteins
Z, Y, A (β-galactosidase, permease, acetylase)
that handle the sugar lactose. Now add lactose
to the broth, and suddenly the three proteins begin to be made at rapid and comparable rates, proportional to the rate of bacterial growth. That
is, the three proteins in question are synthesized

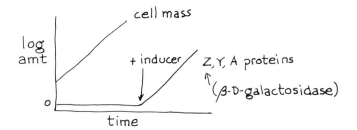

only when the cell needs them. This process is
called enzyme induction and is observed for many
other substances besides lactose and for all bacteria. In each case, specific genes respond to
specific inducing substances.

A completely analogous phenomenon is enzyme
repression. A culture of E. coli in a minimal
medium makes, for example, all the 9 enzymes needed
to synthesize histidine. If we now add some
histidine to the medium, the synthesis of all the
enzymes needed to make histidine stops very
abruptly.

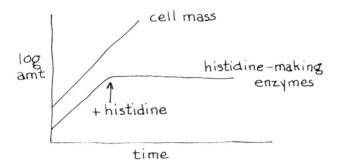

[Note that this phenomenon is different from our old friend feedback inhibition, which is an inhibition of a given enzyme by its ultimate end product. Feedback inhibition is a quick mechanism that serves to save substrate; repression is a mechanism for saving protein. Can you see why both are needed for quick and lasting efficiency?]

Both induction and repression are precise regulatory mechanisms. If we add graded amounts of lactose to the same medium, the enzymes are made at various rates up to a maximum. At that point the genes are functioning at maximum rate, that is, making as many copies of the messenger RNA per unit time as can be made. If the lactose is washed off, the enzymes Z, Y and A are still formed for 1 to 2 minutes--the time it takes before the corresponding mRNA goes to pieces; then no more is made.

What is the molecular mechanism of induction and of repression? The answer comes from genetic experiments.

1. Z, Y, and A are made by three adjacent genes z, y, and a, shown to be different by complementation tests, but all three regulated in a <u>coordinate</u> fashion; that is, if z is 30 percent repressed the other two genes are also working at 70 percent of maximum capacity.

2. There are mutations, called i^-, that cause the three proteins Z, Y, A to be made <u>constitutive-ly</u>, that is, all the time at maximum rate, even when no lactose is present (the i^- mutation affects only these three proteins, that is, it is specific for that group of genes).

3. There is a different set of mutations, called o^c, that also cause constitutive synthesis of Z, Y, and A.

4. There are mutations, called p^r, that reduce the maximum rate at which a cell can make Z, Y,

and A.

   5. The map of the corresponding genes or mutant loci is:

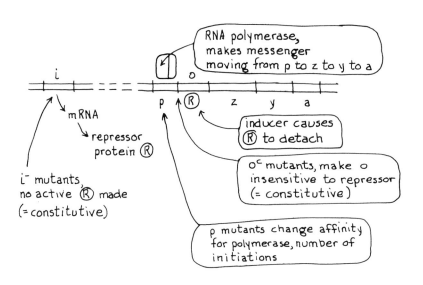

It is an attractive (and valid) hypothesis that a single mRNA molecule has the information for Z, Y, and A--a <u>polycistronic messenger</u>--and that i, p, o are regulatory sites. To test this we use complementation, that is, we construct diploid bacteria by the F' method. The findings

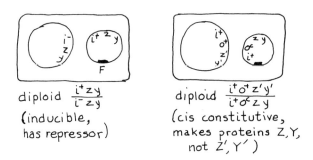

are as follows:

   1. Diploids i^+/i^- are inducible, that is, they make enzymes only in response to lactose or its analogues.

   2. Diploids o^+/o^C are constitutive, that is, they always make enzymes. <u>But:</u> if a diploid is

$o^+z^+y^-/o^Cz^-y^+$, it makes constitutively only the Y protein, which is <u>cis</u> with o^C, not the Z protein that is <u>trans</u> to o^C. That is, i^+ is trans dominant, o^C is <u>cis dominant</u>. Likewise, a diploid $o^+z^-y^+/o^Cz^+y^-$ makes only z^+ constitutively.

3. Finally, p^+/p^r makes reduced amounts only of those proteins whose genes are cis to p^r.

The model that explains all this is shown above. Gene i produces a specific <u>repressor</u> I, which attaches to the "lac region" DNA at the site o (<u>operator</u>) and prevents transcription, that is, prevents RNA polymerase from initiating mRNA synthesis. Lactose or other inducers combine with I causing it to come off the operator. The promoter p is where RNA polymerase attaches to the DNA; then it must slide along the operator region to get to the genes z, y, and a. A mutation in p reduces the efficiency of attachment of RNA polymerase.

Operon

This <u>operon</u> model has been verified experimentally in all sorts of ways. The most dramatically convincing was the use of the so-called Zubay system: an extract of <u>E. coli</u> is made in which enzymes, ribosomes, and other components are so well preserved that addition of a DNA causes the production of the corresponding messenger and also of the proteins coded for by the messenger. In the Zubay system, DNA containing the z, y, and a genes initiates production of the corresponding three proteins. If lac repressor is added, lac messenger and lac enzymes are not made. If lactose is then added, the repressor becomes inactive and synthesis of messenger and enzymes resumes. By physical tests on DNA one can actually show that the repressor protein combines with DNA at the right place and comes off if lactose is added.

[As an exercise, work out a scheme to explain how, for example, histidine causes repression of its synthetic enzymes by activating a repressor (rather than inactivating it). Also, figure out a different case, in which the sugar arabinose induces production of the enzymes that utilize it, but does so not by inactivating a repressor but by activating an inducing protein.]

Biologically speaking, the fundamental point is that in these regulatory mechanisms natural selection has perfected a remarkable class of regulatory proteins. These, on the one hand,

recognize and block (or unblock) specific
stretches of nucleotide pairs in DNA. On the
other hand, they recognize specific effectors--
lactose, histidine, arabinose, or their deriva-
tives--which modulate the interaction between
regulator protein and DNA sequence. Once more we
encounter the basic fact that perfecting of proteins for specific tasks has been the essential
achievement of evolution. Yet it suffices of one
mutation in the i gene, for example, and regulation
of a set of genes is lost. You must never enter-
tain the illusion, common among naive biochemists
and others, that the organisms that exist have been
"perfected" by natural selection for optimal function. Natural selection does not seek perfection:
it simply makes the best it can for those organisms
that managed to survive--by the skin of their
teeth.

Further reading   Stent's Chapter 10, "Conjugation," supplements
this lecture well.

Lecture 18

Phage cycle

At this point you know almost everything important about the molecular biology of the genes. You are almost ready to pass from E. coli to the next most important organism, Homo sapiens. But there remain a few things that only microorganisms can teach us. We shall, therefore, make a short detour du coté de chez T quatre. [If the reference escapes you see M. Proust, A la Recherche du Temps Perdu.]

You already know that bacteriophages (or phages for short) are viruses that use bacteria as their hosts. The host provides a mechanism for the virus to grow and is usually destroyed in the process. A phage has a well-defined reproductive cycle within its bacterial host and a genetic system of its own. A phage particle or virion, with its protein shell, attaches to the bacterial wall and injects its DNA into the cell. (We won't bother with RNA phages here.) Once in the cell the phage DNA can do one of two things. The

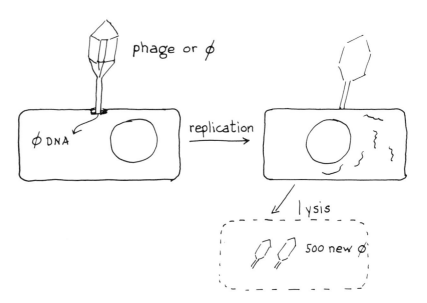

phage or ∅

∅ DNA

replication

lysis

500 new ∅

DNA of certain phage types can behave somewhat like an F factor, that is, it can form a closed circle and integrate itself into the chromosome by crossing-over. It then remains in the chromosome for many generations, as if it were a group of bacterial genes. Occasionally this phage DNA comes out of the chromosome again and produces many phage virions. When it does so, the phage DNA may take along a few bacterial genes. These transducing phages can be used like F' factors, to experiment with partial diploids.

The other alternative is that the phage DNA, once injected into the cell, replicates and produces many copies of itself. It also directs the formation of many phage-specific proteins, which are ultimately assembled with the phage DNA to form new phage particles. These then become free of the bacterium because the bacterial wall is dissolved by a phage-specific lysozyme (similar to the egg-white enzyme whose structure I described earlier in these lectures). The cell <u>lyses</u> and may release as many as 500 or more virions.

If some phage particles are carefully spread on a nutrient agar plate together with sensitive bacteria, next day one can see in the bacterial layer holes or <u>plaques</u> which are of phage. A phage has entered a bacterium, has lysed it, then its descendants have done the same with neighboring bacteria and so on till growth stops for lack of food. We can "count" phage, therefore,

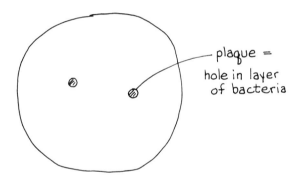

by the number of plaques just as we count bacteria by the number of colonies.

Many phages, besides having genes for the proteins of their shell, also have genes coding for certain enzymes that the phage needs, for example, for replicating its DNA. We know these phage genes because, like the genes of all other organisms, they mutate. There are phage mutants that make bigger plaques, or smaller plaques; some mutants do not grow at all in some bacterial strains but can still grow on others. This is, in fact, how geneticists first learned of nonsense mutations that produced nonsense codons. Some phage mutants were discovered that grew rather poorly in certain <u>E</u>. <u>coli</u> mutants and not at all on the normal <u>E</u>. <u>coli</u> strain:

E. coli

	B	Bx
T4	+	+
T4m	−	+

T4 — CAA

T4m — UAA

nonsense

The phage had a nonsense mutation in one of its
essential genes. It could not grow in the regular
E. coli cells, because this nonsense mutation
caused premature termination of the essential protein. But it grew on the mutant cells because
these had a suppressor tRNA, which placed a usable
amino acid in the place specified by the nonsense
codon.

Recombination in phage

The phage we want to learn about is T4. It is
a big phage, with about 80 already known genes (its
DNA is about 1/15 the size of the E. coli DNA).
If sensitive bacteria are mixed with two different
mutants a and b of T4, the progeny phage contains
four phage types: a, b, ab, and a^+b^+. There is
genetic recombination, just as in Neurospora or in

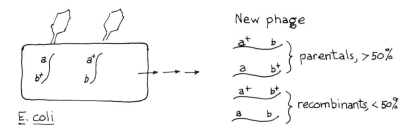

New phage

a^+ b } parentals, >50%
a b^+ }

a^+ b^+ } recombinants, <50%
a b }

E. coli

E. coli. We should use the proper genetic sym-
bolism and write the four types as: a^-b^+, a^+b^-,
a^-b^-, a^+b^+. Genetic recombination, as usual,
serves to make maps. The genetic map of T4, like
that of E. coli, is a single circular chromosome.
Genes that control various proteins of the phage--
head proteins, tail proteins, enzymes that the
phage causes to be made--are mappable at specific
locations.

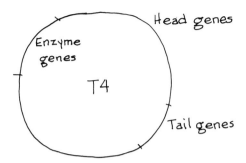

Sex, for phage T4, simply means that two (or more) T4 particles that differ by some mutations enter the same bacterium. As the phage DNA replicates, DNA molecules come together and give rise to recombinants according to the usual rules, as in bacterial recombination. This kind of sex makes it possible for the sex-minded investigators to observe some of the fine points of gene behavior.

Suppose you have two T4 mutants, neither of which can grow in the cells of a given bacterial strain (no phage produced, no cell lysis). You mix these cells with a mixture of both phage mutants and find that the cells lyse and phage is produced. What comes out is a mixture of the two mutants. How did phage manage to grow? The answer is well known to you: it is complementation. The two mutations were in different phage genes. Once inside the cell, the functioning gene (say a^+) of one phage complemented the inactive gene (say a^-) of the other and vice versa (b^+ complemented b^-).

[Make sure you understand the difference between recombination, which is the formation of new combinations of genetic material by exchange of parts, and complementation, which is the functional supplementation between two sets of genes. Of course, the two phenomena can both occur. Thus, when there is complementation between two mutant phages, ab^{++} and a^+b, the progeny phage contains mostly a^+b and ab^+, but also some ab and a^+b^+ recombinants.]

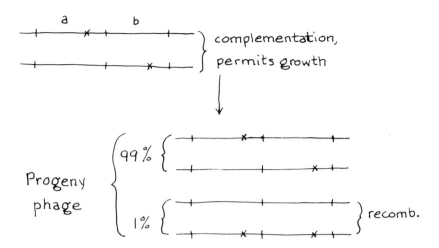

Progeny phage

99% recomb.

1%

recomb.

T4 rII mutants

     There are in phage T4 two wonderful genes, called rIIA and rIIB. They were introduced into molecular biology by Seymour Benzer, a solid-state physicist who turned into one of the most elegant experimenters among geneticists. Genes rIIA and rIIB are immediately adjacent to one another. Both genes in their normal state are needed for the phage to grow on a bacterium called K(λ), or K for short, but they are not needed for phage to grow in another bacterium called B. It is fairly easy to isolate several hundred different rII mutants, because they produce a bigger plaque on B bacteria. Then by recombination tests one can map them. That is, one takes pairs of rII mutants, mixes them with B cells, takes the phage that comes out and measures the frequency of re-combinants rII$^+$. These are easy to count since they can grow on K cells. By this test, all rII mutants fall into one narrow segment of the T4 map. If they are tested again in pairs, but this time for their ability to complement each other-- that is, by mixing two different rII mutants with cells of strain K to find out which pairs manage to grow in K cells--all mutants can be classified into two complementation groups, rIIA and rIIB. All rIIA mutations map to the left of all rIIB mutations.

<u>E. coli</u>

	B	K	
T4 r+	+	+	
T4 rII	+	−	
rIIA and rIIB	+	+	(complementation)

We must, therefore, consider any rII mutant as
being either rIIA⁻rIIB⁺ or rIIA⁺rIIB⁻, the minus
sign meaning the nonfunctional gene. One copy of
rIIA⁺ and one copy of rIIB⁺ in the same bacterium
make phage growth possible.

Since A⁻ phages cannot grow on K at all but A⁺
recombinants can grow (and likewise for B⁻ and
B⁺), this system allows one to detect recombinants
down to extremely low frequencies. Remember the
concept that there must exist a quantum of genetic
recombination corresponding to recombination between two adjacent nucleotides. The frequency of
recombination within the rII genes turns out to
be about 1 in 10^4 recombinants. And, in fact,
rIIA and rIIB genes together comprise about 10^4
nucleotide pairs.

Deletions

One often finds among rII mutants some that
fail to recombine with a group of other mutants.
Usually two mutations that do not recombine are
considered to be changes at the same site of the
DNA. But how can a mutation be at more than one
site? What has occurred is a <u>deletion mutation</u>,
that is, the loss of a piece of the DNA. A deletion mutant fails to recombine with any mutation
located at a site within the piece of DNA that was
lost by deletion.

We distinguish, therefore, <u>point mutations</u>, which
are changes at one genetic site, usually affecting one nucleotide, from <u>deletion mutations</u>, which
are losses of one or many nucleotides. (There
exist also <u>addition mutations</u>, when one or more
nucleotides have been added. These will enter
the picture shortly.)

Deletion mutations make possible a fancy way of
mapping new mutations within a given gene. If we
have a set of ordered deletions whose ends are
known from previous mapping, we can locate the
site of a new mutation by finding, in a series of
crosses, which deletion mutants give recombinants
with our new mutant. This method, applied for
example, to the rII system of phage T4, makes

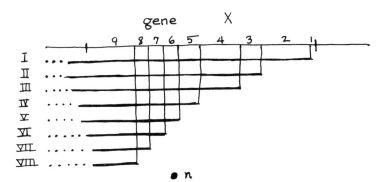

Deletion mutants I to VIII define 9 regions within
gene X. A new mutation n is mapped in region 6
because in pairwise crosses it gives some X⁺
recombinants with deletion mutants VI to VIII
but not with I to V

feasible some very complicated experiments that
require the use of many mutations within a small
segment of a gene.

Sidney Brenner and Francis Crick have used the
rII system in an elegant analysis of how the
genetic code is organized. Their experiments were
done before the actual code was known, and they
used purely genetic techniques made possible by
the experimental flexibilities of the rII system.

It seemed a priori reasonable that the code
was a triplet code. It also seemed reasonable
that it should be comma-free, that is, no nucleotides were used to mark the intervals between
codons. Brenner and Crick, therefore, postulated
that the RNA messenger that coded for a given protein was translated from a fixed point in a single
direction, each amino acid being represented by
three nucleotides. This would be possible if the
translation from RNA to protein were done according to a "reading frame," which shifted along the
RNA three nucleotides at a time. We now know, of
course, that the object that moves along translating the RNA three-by-three is the ribosome.

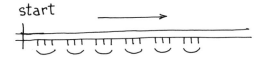

Brenner and Crick's experiments were based on
the observation by Benzer that the beginning

section of the rIIB gene seemed to be useless: a
deletion that removed that part of the gene did
not destroy the ability of the phage to grow on K.
And yet, mutations <u>within</u> that useless section
did prevent the phage from growing on K cells.
(You can easily guess why. A gene codes for a
polypeptide chain. A mutation may eliminate an
end of the chain without making the protein inactive. A change of amino acid within the useless
portion may prevent the proper folding of the
chain into a functional protein.)

Frameshift mutations   Within the useless section of the rIIB gene
Brenner and Crick proceeded to isolate a series
of special rIIB⁻ mutants. They used a mutagen,
the dye acridine orange, which is known to inter-
pose its ring-shaped structure between the base
pairs of DNA. DNA replication in presence of
this dye produces mostly mutations of the "add"
or "subtract" types; that is, mutants with one
extra nucleotide or one missing nucleotide.
Finally, they confirmed the identity of their
mutations as "add" or "subtract" (plus and minus)
by recombination mapping.

A mutant that has either a plus or a minus mutation is hopeless: all triplets beyond the mutation are wrong (wrong amino acids, that is) and
the entire protein chain is "gibberish." The
reading frame is shifted off phase. This is why
these mutations are called <u>frameshifts</u>. A recom-

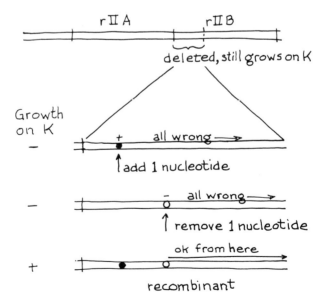

binant that contains one "add" and one "subtract"
may have a functional rIIB gene, however, because
recombination has restored the reading frame in
the significant part of the gene. (Note the
critical requirement that the sequence between
the two mutant sites--a plus and a minus--does
not include (1) any amino acids that may interfere with the assembly and/or function of the protein or (2) a triplet corresponding to a termination signal, i.e., UAA, UAG, or UGG on the messenger.)

Having reconstructed a functional gene by combining a deletion and an addition, they could go
one step forward. They brought together by recombination three mutations of the same type, that
is, three pluses or three minuses. If these combinations could restore the function of gene rIIB,
they could conclude that the translating frame
read the code by triplets. This was exactly the
experimental result. Any combinations of 2 or 4,
or 5, mutations failed to generate a functional
rIIB gene. Only groups of 3 plus or 3 minus could
work.

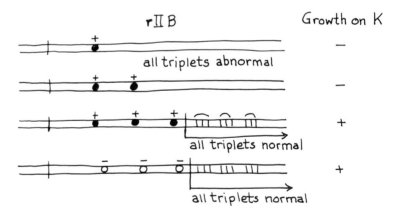

Unfortunately, the proteins made by the rIIA
or rIIB genes were not known. The beautiful
genetic work remained without the validation that
had to come from the analysis of the proteins
corresponding to the recombined genes. Luckily,
there was another gene easier to work with. This
was the gene of T4 that makes the lysozyme that
dissolves the host cells. Streisinger and Tsugita
knew the amino acid sequence of this lysozyme.
So they proceeded to repeat the plus and minus
experiments on the lysozyme gene. Taking a phage,

in which a plus and a minus mutation very close
to each other had reestablished the lysozyme function, they purified the lysozyme and compared it
to that of the normal phage. They found in the
middle of the protein molecule a stretch of five
amino acids that were changed. And the changes
were exactly those predicted by the genetic code
(which by then was known) for the triplets present
between a one-nucleotide deletion and a one-
nucleotide insertion.

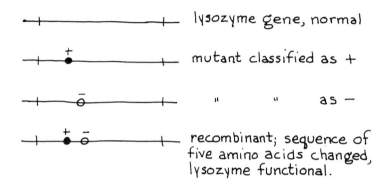

lysozyme gene, normal

mutant classified as +

"   "   as —

recombinant; sequence of
five amino acids changed,
lysozyme functional.

    With this illustration of the power of genetic
analysis we leave the more specifically molecular
aspects of genetics. We shall turn next to
aspects of genetics that concern other organisms
than bacteria or phage. We shall see that in
complex organisms, with elaborate developmental
processes, a major role of genes is the timing
of steps in temporal sequences of events. This
will lead us into development and physiology,
which are the expression of the genetic program
in terms of specialization. What you have learned
of genes, gene function, gene regulation, and gene
organization should be kept in mind as the basis
of the entire fabric of biology, from evolution
to the functioning of the brain.

Further reading   Stent's Chapter 11, "Phage Growth," goes more
deeply into one aspect of this lecture. So do
Watson's Chapters 7, "The Arrangement of Genes on
Chromosomes," and 17, "Regulation of Protein
Synthesis and Function."

Lecture 19

Eucaryotic genetics

At this point we have learned about the nature and the mode of action of the program of an organism, at least of a procaryotic organism such as E. coli. Our task, in the rest of these lectures, is to "construct" a complex, multicellular organism on the basis of its own program.

As a start, let us concentrate on the genetics of complex organisms, including Homo sapiens, which obviously has a special interest for us. Keeping in mind what we know of the organization and function of the program in simpler organisms, we must ask what new feature complexity adds to the picture.

Chromosomes

Man's body cells have 46 chromosomes, properly replicated and distributed at mitosis and meiosis and containing about 1,000 times as much DNA as a cell of E. coli does. Cells of Drosophila melanogaster, the fruit fly, have 8 chromosomes. The

XX   XX XX   XX XX

XX XX XX XX XX XX XX

XX XX XX   XX XX XX

XX XX XX XX XX (ʌ)
              X   Y

Human chromosomes

numbers and organization of chromosomes are quite stable for a given species, but can vary even among related species. For example, in the genus Drosophila there is a variety of species, with different arrangements of the same genetic material into different chromosome numbers. These have come into being because chromosomes undergo several types of rearrangements: deletions, inversions, or translocation of pieces. Although rearrangements can be harmful if they damage useful genes, sometimes they create new and favorable gene combinations and can play a big role in evolution.

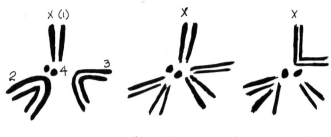

D. melanogaster
4 pairs

D. virilis
6 pairs

D. pseudo obscura
5 pairs

Chromosomes are thicker and more complicated objects than DNA fibers. Even when they first become visible in the prophase of mitosis or meiosis (they are not seen in the interphase period), they are almost 1 µm thick (remember that a DNA fiber is 2 nm thick). In chromosomes, DNA is associated with several classes of proteins, including an important class of basic proteins (that is, proteins with many lysine or arginine side chains) called <u>histones</u>. Histones have remained almost unchanged in the course of evolution; it is now believed--without any really strong evidence--that they play some regulatory role in turning genes on and off in eucaryotic cells.

Chromosomes and DNA

How is DNA, as much as one centimeter in length, arranged within a chromosome 1 to 10 µm long? Is it in a single piece or many? The evidence, although not one hundred percent solid, by-and-large supports the idea of a single DNA fiber per chromosome (or, rather, two fibers, since each chromosome when it appears at mitotic prophase consists of two <u>chromatids</u>, which then separate).

First of all, the genetic map of all chromosomes is always linear. The genetic map of <u>Drosophila</u> consists of 4 linear gene sets, as many as there are chromosomes.

66 recombination units _____ X chromosome

107 units _____ 2 nd

106 units _____ 3 rd

0.1 u 4 th

The "genetic length" of each set is roughly
proportional to the length of a chromosome.
(There is no map of the Y chromosome because in
Drosophila the males happen to have no crossing-
over, and at any rate there is nothing for Y to
pair with.) Since each gene set is linear,
either there is only one DNA fiber or, if there
are many, they must behave genetically as if
they were in a unique unchanging linear sequence.
In some insects that have visibly banded chromosomes, the genetic map matches precisely the
visible map as revealed by chromosomal rearrangements. The DNA molecules extracted from animal
cells are often long enough to suffice for an
entire chromosome.

A beautiful autoradiographic experiment by
Taylor strongly favors the single DNA fiber
theory. Cells were labeled with ^{3}H thymine while
they were making DNA (the S period of the cell
cycle) and then were studied in the presence of
the drug colchicine. Under these conditions, the
chromosomes remain in the same cell for several
cycles of replication, because there is no mitosis
or cell division. The results fit nicely the idea
that each chromatid behaves as if it contained
just one DNA double helix, with semiconservative
replication at each S period of the cell cycle
(and occasionally some "sister-strand" exchange).

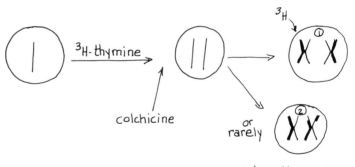

① Semiconservative replication
② Sister-strand exchange

In the chromosome, of course, the DNA fiber
must be all balled up or curled up like string in
a paper bag, although each segment of it must be
accessible when it functions as template for RNA
synthesis or for replication. Does one ever see

chromosomes in this state? The best example is
seen in meiosis of amphibia. During the maturation of the egg cells, as we shall see later,
there occurs a synthesis of enormous amounts of
RNA. In the maturing egg cells or <u>oocytes</u> one
sees thin symmetrical loops sticking out on each
side of the chromosomes. The gene or genes on

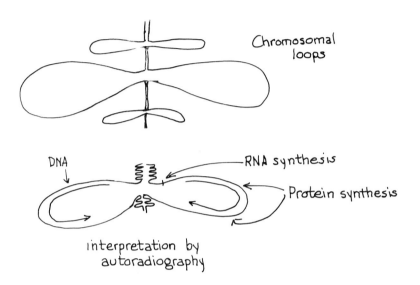

each loop are transcribed to make RNA and then
reenter the main chromatid thread (a bit like
pulling out a thread from a thick woolen cord and
then pulling it back in). By treating oocytes
with x rays, which break DNA molecules, and then
examining them, Gall showed that to cause a break
in a loop two hits were needed, just as two x rays
are needed to break double-stranded DNA fibers in
the test tube. (That is, the number of breaks increases with the square of the x ray dose.)

So this is our tentative picture of a chromosome after it has completed DNA synthesis: two
chromatids, each containing a continuous DNA double
helix, with loops of DNA sticking out where the
genes are functioning. In most chromosomes, of
course, the loops are relatively few and far between and are not readily seen, but in some exceptional cases of giant chromosomes the loops are
visible as "puffs" on the edges of the chromosomes.

Human genetics   Let us come to man. We know our own genetics
less well than that of <u>E. coli</u> or <u>Drosophila</u>. Our
generations last longer, and crosses are (fortunately) not controlled experimentally. We learn

of our own genes by observing hereditary differences and tracing back these differences through pedigrees. Eye color, hair or skin characters, and the ability to curl one's tongue are readily observable traits; some of them are simple, others are very complex in their genetic determination. (The inability to curl one's tongue, as I mentioned in an earlier lecture, is due to a recessive gene that is present in 40 percent of our population. Can you calculate your chances of being a noncurler?)

Heredity and disease

The most interesting genetic traits in man are those that result in disease. A disease need not be <u>congenital</u>, that is, present at birth, to be hereditary. Hereditary diseases occur at various times in life because the genes involved become functional at different times in life. Some genetic diseases are dominant, others recessive. Some are due to mutations in one gene, others to rearrangements of chromosomes. Some appear to affect just one recognizable trait, others seem to be <u>pleiotropic</u>, that is, to have multiple effects. The important thing to remember is that most genetic diseases are due to the defective action of a single gene or a pair of homologous genes.

Galactosemia

As a first example, take the disease <u>galactosemia</u>. Some babies, if given milk after birth as usual, fail to grow well and soon show symptoms of permanent damage to brain and liver and develop cataracts, that is, opacities in the lens of their eyes. If these babies live, they are forever crippled in body and mind. If, however, these babies are not given milk or are taken off milk almost immediately they can be perfectly normal. All these symptoms come from being homozygous for a recessive gene whose normal, dominant allele makes one of the enzymes needed to use galactose, which is part of milk sugar.

$$\text{Galactose} \xrightarrow{\ K\ } \text{gal-1-phosphate}$$
$$\downarrow T$$
$$\text{UDP-glucose} \xleftarrow{\ E\ } \text{uridine diphospho gal}$$
$$\text{(UDP-gal)}$$

In the absence of the enzyme T the substance

galactose-1-phosphate cannot be used and accumulates in cells, damaging those cells that are
especially sensitive at the time when exposure to
milk occurs. Note the similarity with biochemical
mutations in bacteria: one enzyme is missing,
one pathway is blocked.

For a baby to be galactosemic, both parents
must be heterozygous for the galactosemia gene:
Gg x Gg → 1 GG, 2 Gg, 1 gg. One-fourth of the
offspring of a couple of Gg heterozygotes is expected to be galactosemic. Is it possible to
detect the heterozygous parents and warn them of
the 1-in-4 risk, and therefore advise them not to
nurse or feed milk to their babies? One can do
so by measuring the level of the critical enzyme
in the red blood cells. Despite some overlaps,
the levels average 32 for normals, 16 for heterozygotes, and less than 4 for galactosemics. Any-
one with less than 27 or 28 can be tested more
fully in a specialized clinic. The children of
two heterozygous parents can be normal if they
get no milk. Note that this is a clear example
that a genetic trait is or is not a "defect"
depending on the environment. Nature and nurture
cannot be separated.

Hundreds of defects of human metabolism have
been traced more or less precisely to specific
genes. Many of them are mapped on specific chromosomes. Some of these defects are especially instructive, either physiologically or genetically.

Huntington's chorea   The disease Huntington's chorea is produced by
a dominant gene (that is, one copy of the gene is
enough to produce the disease; the normal allele
is recessive). But its manifestation is late in
life: only after the age of 35 or 40 years does
the affected person begin to show all sorts of
symptoms of brain degeneration. How can a carrier

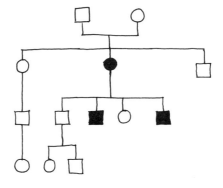

Pedigree for a dominant character

person learn that he has the chorea gene and
refrain from having children, 50 percent of whom
will have the disease? This can be done only by
pedigree analysis. Except in the case of a new
mutation in the germ line, a chorea patient always has one parent with the disease. [Note in
the pedigrees: □ = male;   ○ = female;   ■ or
● means presence of the disease;   □-○   means
a pair with children; ◇³ means 3 children, sex
unknown, without the disease]. This instance
shows the importance of pedigree analysis in human
medical genetics. Not all dominant genes show
their effects in all individuals or on all sus-
ceptible parts of the individual. Geneticists
refer to this as <u>incomplete penetrance</u> of the
gene effects.

Hemophilia   Another example, of a different kind, is <u>hemo-
philia</u>, a defect of blood clotting. (Blood clotting
is a complicated process that leads to the formation of fibrin from fibrinogen, a blood protein.
Fibrin forms the main part of the blood clot.)

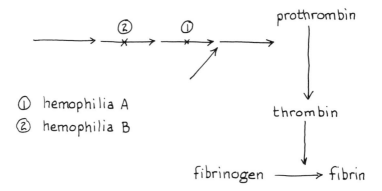

① hemophilia A
② hemophilia B

On this pedigree, there is a remarkable regularity:
only males have the disease and do not transmit it
to their offspring. Females in the pedigree are
never hemophiliac, but can transmit the condition
to their sons. What kind of inheritance is this?

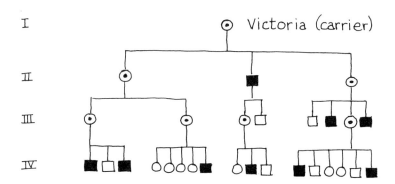

⊙ = heterozygotes

Sex-linked traits   The time has come to talk sex. Sex in mammals
and many other animals is determined by a subset
of the chromosome contingent, called the X and Y
chromosomes. In man and in all mammals, XX is
female, XY is male; Y has genes for maleness; it
is the sexist chromosome. The other chromosomes,
present in both males and females, are called
autosomes. (In Drosophila, males are XY, but
maleness only requires having one X instead of
two, so that an accidental XO is male. In frogs
and fishes and bees there are other arrangements.)
If you look at the hemophilia pedigree, you'll
notice that hemophilia behaves like a recessive
gene in an X chromosome, that is, it is a sex-
linked gene. Females have two X's, therefore a
single hemophilia gene does not show up; but one-
half of the sons of a heterozygous female is
struck by hemophilia. [Can you figure out how a
female can be hemophilic?] There are many sex-
linked genes, because the X chromosome is a large
one. For example, color-blindness and hemophilia
are linked, with about 10 percent crossing-over
in female meiosis.

Queen Victoria of England (and Empress of
India!) was a carrier of hemophilia, due to a
mutation that must have occurred either in the
germ line of her parents or in her own early
development. Since her numerous daughters were
highly prized mates for all sorts of European
princes, she flooded the royal crowd with hemophilia and probably contributed to the downfall
of royalty--a sort of one-gene Watergate. Such a
situation in which an individual can be established
to have initiated a lineage with a new gene is

called the <u>founder effect</u>.

In man as well as in other animals there are many known genes located, like hemophilia, in the X chromosome. In a sense, one might call maleness a disease, although an obviously necessary one, at

Xg   G6PD   CB   Hem A   Hem B

Xg =   a blood group gene

G6PD=   glucose-6-phosphate dehydrogenase

CB   =   color blindness

Hem A= hemophilia A

Hem B= hemophilia B

least in animals. Many plants do without a genetic mechanism for sex and are hermaphrodites: the same plant produces male and female germ cells in flowers or spores.

Sickle cell anemia

Hemophilia, at least its most common type, is due to the absence of a protein needed for proper blood clotting. Again as in galactosemia the defect is the absence or greatly reduced function of a single gene product. Another disease, sickle cell anemia, is a prototype of a class of conditions due to the presence of an abnormal gene product. Hemoglobin S differs from normal hemoglobin A in being $\alpha_2\beta^S{}_2$ instead of $\alpha_2\beta^A{}_2$. The gene for β chains has mutated from the β^A to the β^S form, putting valine instead of glutamic acid at the sixth position from the amino terminal in the chain. Hemoglobin S causes lots of trouble: when oxygen levels decrease, the hemoglobin molecules stack into bundles, the cells "sickle" and clog the capillaries in every organ of the body. The individual is in trouble. In fact, those homozygotes for β^S that reach adulthood are mostly crippled because of extensive damage to many organs when oxygen supply fails. But the heterozygotes also suffer: they have what is called the <u>sickle cell trait</u>, a series of mild disturbances due to poor ability to adapt to low oxygen conditions such as hard exertion.

The β^S gene is neither dominant nor recessive: it is <u>codominant</u>. The heterozygote's red cells have a mixture of the two hemoglobins, but the

presence of β^A does not prevent the manifestation
of β^S. There are many genes of this kind, in
which the abnormal gene product is detected in
heterozygotes. We may ask, how abnormal is an
abnormal gene? About 10 percent of black
Americans have the β^S gene. If it were fully
abnormal, would it not have disappeared? It
turns out, as you probably all have heard, that
this abnormal gene was beneficial in West Africa
because it protected heterozygous carriers against
malaria. The malarial protozoan <u>Plasmodium</u>
<u>falciparum</u>, which must spend part of its life
cycle within red blood cells, does not like hemoglobin S and fails to proliferate!

This example is instructive: in evolution
there is no absolutely good gene or absolutely
bad gene. There are alternative forms of genes,
which arise by mutation. Most mutations produce
genes that function less efficiently or not at
all and therefore cause the carrier individuals
or some of their descendants to be eliminated.
They can persist for many generations, however,
because the mutant allele is usually recessive to
the normal allele of the same gene. Occasionally
such a mutant gene provides some advantage to the
individuals that carry it by improving survival or
fertility. This can happen especially when a
mutated gene finds itself in a new gene combination
that arises by sexual reshuffling. Evolution proceeds blindly but astutely.

Fetal hemoglobin

Hemoglobin can undergo other mutations comparable to the sickle cell one: mutant hemoglobins that are widespread in nature are also probably of some advantage in special environments.
One of the most interesting features, however, is
not mutational but developmental. In mammals,
the hemoglobin of the fetus is different from that
of the adult: hemoglobin F or $\alpha_2\gamma_2$ instead of
hemoglobin A or $\alpha_2\beta_2$. This is not a matter of mutation. The γ chain is made by a different gene
than the one that makes β. In the early fetus
the γ gene is functional and the β gene much less
so. After the eighth month of fetal life less
and less γ chain is produced, and practically none
none is produced by the time a baby is six months
old. Directly or indirectly the β and γ genes
regulate each other's action, or possibly are under
a common regulatory control that turns one off and
the other on. In fact, when the β chain is defective
as in sickle cell anemia, γ chain continues to be

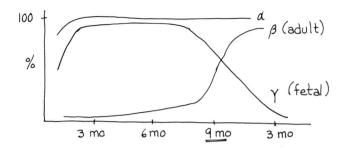

made in adult life! The regulatory mechanism is
not yet known. It may resemble those we learned
about in E. coli. But, whereas the regulation in
E. coli is directed to coping with changes in the
external medium, in man the regulation of hemoglobin genes is needed to express the genetic program in a precise developmental sequence: from
fertilized egg to embryo to fetus to child to
adult. Why should the hemoglobin type change at
birth? No one knows for sure. Hemoglobin F in
the fetal blood does have a somewhat higher
affinity for oxygen than does hemoglobin A in the
adult circulation; but this difference is hardly
enough to explain the switch. This does not mean
that the switch occurs without reason: we believe
that practically all genetic regulatory phenomena
have evolved because of important selection pres-
sures. There is probably some important reason
why hemoglobin F is good for the fetus or alternatively why hemoglobin A is bad for it.

Further reading

Strickberger's Chapter 12, "Sex Determination
and Sex Linkage in Diploids," is useful for further study on sex determination.

Lecture 20

Chromosomal
abnormalities

You saw in the preceding lecture that changes in
chromosomal structure have taken place, for example, in the evolution of the Drosophila flies.
Each species, however, once well established in
nature, has a fixed number and pattern of chromosomes that represents its caryotype. In animals
at least, even the accidental loss of a piece of
chromosome is almost always lethal. Aneuploidy,
that is, the departure from the normal condition
of two copies of each chromosome (euploidy), is
generally not tolerated, even when it involves
added rather than lost genetic material.

Down's syndrome

Children of older mothers have an increased
chance of being "mongoloid," a misnomer for a
syndrome characterized by more or less distorted
features and a mental idiocy due to abnormal brain
development. This is called Down's syndrome.
The caryotype of such an individual often shows a
typical abnormality: 3 copies of chromosome No.
21, either all free or one of them attached to
some other chromosome. Even the presence of an
extra fragment of No. 21 is sufficient to cause
more or less severe Down's syndrome.

Why would the mother's age make any difference?
The egg cells of a woman are all ready at birth
but await sexual maturity before undergoing meiosis, one each month according to the menstrual
cycle. The older the woman, the older are her
eggs, and the greater is the chance of something
going wrong in meiosis. If two No. 21's undergo
nondisjunction, that is, go to the same daughter
cell at one of the two meiotic divisions, or if

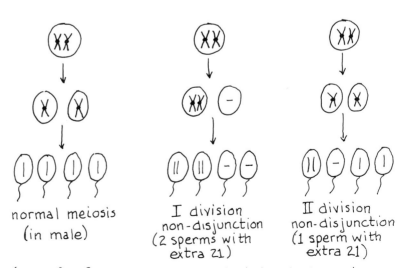

normal meiosis
(in male)

I division
non-disjunction
(2 sperms with
extra 21)

II division
non-disjunction
(1 sperm with
extra 21)

(Same for female meiosis, except that only 1 egg is
formed; the other 3 nuclei get lost)

one No. 21 gets stuck to some other chromosome,
the fertilized egg can get two of them and, with
one 21 from the sperm, this makes a zygote with
three. The reason similar situations are not
found for other chromosomes is probably that they
would be completely lethal, causing very early
abortion.

Which brings us to a practical point. Doctors
can diagnose Down's syndrome by means of amnio-
centesis: that is, by drawing some of the amniotic fluid in which the fetus is suspended and
examining the fetal cells that are present in it.
The test can be performed by an expert by the end
of the third month of pregnancy, late enough to
have enough amniotic fluid and early enough for a
reasonably safe abortion. Pregnant women over 35
years of age are well advised to seek such a test
and, if Down's syndrome is found, may wish to
abort.

Even though mongoloid individuals usually do
not reproduce, there is a chance for this syndrome
to be transmitted hereditarily. For example, in
a cell of the germ line a piece of chromosome 21
may become attached to chromosome 15 ("chromosome
15^{21}") while the other piece remains as a 21
fragment. At meiosis there are four types of eggs:
21 + 15; fragment + 15; 21 + 15^{21}; and fragment +
15^{21}. The first is normal; the second is lethal
(too little 21 material); the third, if fertilized,

Down's fragment

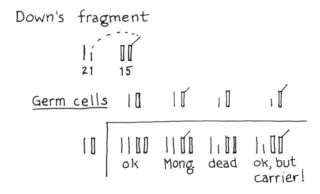

gives a mongoloid; the fourth makes a normal baby,
because it has a full complement of 21 and 15, but
this baby later, when grown up, can produce mongoloid children by making eggs or sperms with
15^{21} + 21. The Down's syndrome story is an excellent illustration of the precision achieved by
natural selection in regulating the exact amount

of gene material that an organism such as man must have.

The sex chromosomes seem to have a bit more leeway to depart from euploidy. This is not surprising since they already have to accommodate themselves to function as XX or XY in the two sexes. At the meiotic divisions the sex chromosomes occasionally get stuck or fail to separate. Thus, besides the normal gametes with X or Y, one can have gametes that contain XX, or XY, or YY, or O (no sex chromosome). These gametes give rise to individuals who are generally viable, although they often are abnormal. XO individuals have Turner's syndrome: they are females, short, sterile, mentally retarded. XXX are females, either sterile or normal. XXY or XXXY are males, sterile, short-limbed (Klinefelter's syndrome). Finally XYY are sexually normal males, usually taller than average.

The effects of these abnormalities of sex chromosome on fertility are worth some attention. In the sterile types the sex glands or gonads-- ovaries or testes--do not develop. That means that a precise dosage of the products derived from genes in the sex chromosomes is needed at critical times of development in order to determine the formation of the gonads. Again we observe the importance, not only of gene action, but of precise gene action at a precise time in development.

A question must already have come to your mind: if the X chromosome has genes other than those for sex characters, for example, the gene whose mutation produces hemophilia, how come males and females are not very different in these traits? The current interpretation, proposed some years ago by Mary Lyon, is that in the course of embryonal development of a female either one of the two X chromosomes becomes permanently inactivated. The inactivation occurs in various cells at random and is irreversible, not only in the cell where it occurs but in all its descendants. Each woman is a mosaic for her two X chromosomes, one derived from the mother, the other from the father. In each resting cell of an adult woman or other female mammal, one sees microscopically a blob of chromosomal material, called the Barr body, all condensed and sad looking, which corresponds to the inactive X chromosome.

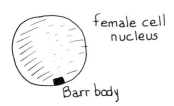

female cell
nucleus

Barr body

Strong evidence for the correctness of the theory has come from women who are heterozygous for two alleles of a sex-linked gene that makes a well-known enzyme, G-6-PD (or glucose-6-phosphate dehydrogenase). If one grows cultures from single cells derived from such heterozygous women, one finds that the cells of each culture have only one or the other of the two forms of the enzyme, never both. The X chromosome that had become inactive during development does not return to activity even in cell culture: it is irreparably nonfunctional. (Interestingly enough, the random inhibition of either X chromosome is characteristic of common mammals only. In marsupials, such as the kangaroo, it is always the X from the male parent that becomes inactivated. Can you suggest a mechanism?)

It is clear that in evolving sexuality nature has encountered some conflicting problems. Both X's must function early in development in order to make a normal female since XO females have no ovaries and are sterile; but soon thereafter one of the two X's must become inactivated in each individual cell to make the rest of the body be normal. What kind of signals may initiate this process? How could you try to find out?

(What about hemophilia? If one X does not function, half of the cells of the heterozygous females lack the normal gene. Why are the females not hemophilic?)

Quantitative inheritance

Up to this point we have dealt with clear-cut genetic differences. Either you are hemophilic or you are not; either you have Huntington's chorea or you do not, even though you may have to live and wait till you are 40 or 50 before you know. But there is a class of inherited traits that are not so clear-cut. For example, the height of human beings is clearly influenced by gene action, not only because of the differences in average height between females and males, but because of statistically significant correlations in height between parents and offspring. But the situation is more complex than that: there are many women taller than many men, and numerous tall couples happen to have short children. These quantitative characters exhibit a more or less continuous variation because they are determined by many genes, each contributing to the final result. Quantitative hereditary traits are also subject to influence from the environment, of course: people of Japanese origin raised in the United States, for example, are taller on the average than those raised in Japan, presumably because of dietary differences. But in both groups the range of individual heights is genetically determined.

How can we unscramble the heredity of these quantitative characters? Soon after the turn of the century, the Danish plant geneticist Johannsen coined the word gene to explain the following experiment with beans. The individual beans from a given garden variety weighed from 15 to 90 centigrams with an average around 50 cg. From 19 individual beans Johannsen grew 19 pure lines by repeated self-fertilization. Each of these lines had a characteristic bean weight average, the averages ranging from 35 to 64 with a much narrower spread than the original garden variety. Whether he took the heaviest beans from the lightest variety or vice versa, the average at the next generation remained the same. The explana-

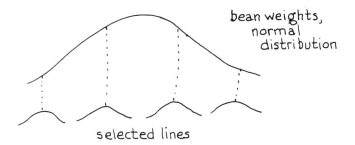

bean weights, normal distribution

selected lines

tion is simple: there are many genes that contribute to bean weight (and likewise to human height). Mendel's laws predict this:

A, B... contribute to weight; a, b... don't

(1)
$$P \qquad AA \times aa$$
$$F_1 \qquad Aa$$
$$F_2 \qquad AA \quad Aa \quad aa$$
Ratios: 1   2   1   coefficients of $(x+y)^2$

---

(2)
$$P \qquad AABB \times aabb$$
$$F_2 \qquad AABB, AaBB, \ldots aabb$$
Ratios: 1   4   6   4   1   coefficients of $(x+y)^4$

There are many genes whose allelic forms influence the same quantitative trait. The frequency of a given class of height or weight depends on the number of relevant alleles present.
If there is no dominance, all alleles of each gene
being codominant, the distribution of frequencies
is symmetrical and approaches a normal distribution. If there is dominance, the distribution is
skewed, more so the fewer the genes that are involved.

Heritability
In most cases a quantitative trait is influenced
both by heredity and by the environment, as we
have seen for body height. That is, the phenotype
reflects both genotype and external conditions.
Geneticists analyze the various components of
the variation of a trait in terms of the variance
V, which is the sum of the squared differences
between the individual values X_i and the mean
value $\overline{X}$, divided by the number N of individuals
minus one:

$$V = \frac{\Sigma_{i=1}^{N} (x_i - \overline{X})^2}{N - 1} \, .$$

The phenotypic variance V_p is composed of three
terms: $V_E + V_G + V_{GE}$, representing, respectively,
the environmental component, the genetic component,
and the interaction between the two. (An example
of interaction would be if a given genetic influence were expressed more or less effectively in
different environments.) One can then define
heritability H^2 as approximately V_G/V_P, that is,
the fraction of variance due to heredity. Controlled tests on plant and animals make it possible
to measure H^2. For example, in Drosophila H^2 is
52 percent for wing length and 18 percent for egg
production, meaning that the number of eggs laid
is much more influenced by food, etc., than the
length of the wings.
The inheritance of quantitative characters
plays a very important role in practical problems,
from growing beans or chickens to school education.
You can select bigger and bigger beans or chickens
for breeding and arrive at a variety whose mean
weight is as high as the species can produce.
Since you are throwing away genes that make for
small size, you also get greater uniformity of
size, as in Johannsen's experiment. You must be
careful, in the process, not to throw away genes

that make for resistance to disease or for softer or tastier meat or for numbers of eggs laid.

A number of observations have suggested that human intelligence, which is an obviously variable trait, has a multigene hereditary component. (One way to estimate heritability in human populations is to compare identical twins, who have the same genes, with fraternal twins of the same sex, who are born and raised together but differ in many genes.) But the study of heritability of intelligence meets the problem of defining intelligence. Educational psychologists have devised the I.Q. tests, which classify individuals in a normal distribution and predict fairly well their performance in school--more specifically, in schools of Western, white, upper-middle-class type; probably not in the Confucian schools of Imperial China or in the Eskimo training for survival on the ice-pack. No serious psychologist has claimed to know what I.Q. actually measures in biological terms. The I.Q. of black children in the U.S. has had in recent years a distribution with a mean about 15 points below the mean of the distribution for white children. The two distributions, of course, are widely overlapping. It is possible, and in fact almost certain, that intellectual qualities, like all other qualities, include a genetic component, but there is no reason to believe that the I.Q. measurements are a valid measure of intelligence or educability. Environmental factors are likely to play the overwhelming role in deciding the way any group of individuals scores in a test like I.Q.

A concept that is highlighted by studies on quantitative inheritance is that of genetic polymorphism. As you already know, within any natural population such as mankind many genes are present in several allelic forms rather than in a unique standard form. Hemoglobin is a good example: sickle cell trait people are $\beta^A\beta^S$, heterozygous for the β chain gene. For genes that determine easily analyzed enzymes, polymorphism is evidenced by the presence of isoenzymes, that is, enzymes produced by nonidentical alleles of the same gene. For example, for the enzyme peptidase A there may be present the diploid constitutions α^1/α^1, α^1/α^2, or α^2/α^2. Since the enzyme is a dimer, its structure can then be:

It has been estimated that a human being is
probably heterozygous for at least 10 percent of
its genes. Even two brothers may differ by 10
percent of their genes or more. Thus the concept
of an ideal prototype for a species, such as is
represented in museums, is a figment of the imagination. Each species, defined as a set of potentially interbreeding organisms, represents a complex of intersecting subsets of genetic endowments.

If one considers that man has at least 100,000
genes, that 10,000 of these may be heterozygous,
and that reshuffling occurs at each generation,
it is evident that each individual is genetically
unique (except for identical twins) and is as much
a representative of the species as any other individual. An African pygmy, an Australian aborigine,
a Norwegian Viking, or an American Indian may provide an equally valid, or rather equally meaningless museum prototype of the species Homo sapiens.

Further reading   Strickberger's Chapters 14, "Quantitative Inheritance," and 15, "Analysis of Quantitative
Characters," are very technical and informative.

Lecture 21

Genetic polymorphism within a species raises the question of <u>gene frequencies</u>. In general, each individual has two copies of each of the genes characteristic of the species (except for the sex-linked genes). These two copies can be different from each other and different from those present in some other individual. Within a population, each allelic form of any one gene is present in a certain frequency, which depends on several factors that we must now explore.

Let us start with PTC or phenylthiocarbamide. This is a substance that to some people tastes horribly bitter, while to others it has no taste at all. "Tasters" have either one or two dominant genes for tasting (possibly the dominant gene T makes an enzyme needed to process PTC in the taste buds). If I were to pass around PTC paper to a class of 160, say, I may find about 108 tasters and 52 nontasters. (In China, I would expect about 140 and 20.)

These are frequencies of the <u>phenotypes</u> in the populations. How do we calculate the relative frequencies of the two genes T and t in the population? Let us call these frequencies p and q. By definition, p + q = 1 if these are the only two alleles. Remember that nontasters are tt; tasters are either TT or tT. The chance of a person being tt is q^2 (chance of getting one t gene from both parents). If the nontasters are $q^2 = 52/160 = 0.32$, the frequency q of t gene is $\sqrt{0.32} = 0.565$. Therefore, the frequency p of T genes is 0.435.

Hardy-Weinberg
equilibrium

For the three types TT, Tt, and tt we have the equation

$$p^2 + 2pq + q^2 = 1 \ ,$$

where p^2 is the frequency of TT's; 2pq the frequency of Tt's; and q^2 that of tt's (nontasters) in the population. For the PTC genes, there are 0.32 nontasters, 0.19 homozygous tasters, and 0.49 heterozygotes in our classroom population sample.

This equation seems almost silly, not interesting enough to deserve its name of <u>Hardy-Weinberg equilibrium</u> equation. Its importance is that it expresses the fact, that, so long as there is no selection for or against either one of the two alleles, the gene frequencies in a random mating population remain constant from generation to generation. You can verify this for PTC tasting

by simply noticing that at the next generation
the frequency of T gametes is again p and that of
t gametes again q = 1 - p:

T genes = 2 x p^2 + 1 x 2pq = 0.38 + 0.49 = 0.87

t genes = 2 x q^2 + 1 x 2pq = 0.64 + 0.49 = 1.13

$\qquad$ 0.87 + 1.13 = 2; 0.87/2 = 0.435; 1.13/2

$\qquad$ = 0.565.

By taking one-by-one the assumptions made in
arriving at this equilibrium we can learn a lot
about populations and natural selection.

First consider the assumption of random mating,
that is, no preferential mating with respect to
the trait in question. The contrary of it is
assortative mating: for example, TPC tasters
might prefer tasters, or vice versa. The gene
frequencies remain constant from generation to
generation [try proving it!] but there are different proportions of homozygotes or heterozygotes.
For example, if tasters started mating only with
tasters, the proportion of Tt would go down by
half in each generation.

Selection

Assortative mating is less important in natural
populations than selection. Suppose that gene T
had a negative effect on reproductive success, so
that TT or Tt individuals had fewer offspring.
Then the gene frequencies would change, tending to
eliminate the T genes and to create a population
of nontasters. It is probably safe to assume
that almost no gene is absolutely neutral: every
allele that exists in a population does confer
some advantage or disadvantage and therefore tends
to be favored or to be eliminated. When frequencies are stable, there must be some advantage
to compensate for any potential disadvantage.
For example, the sickle cell gene β^S reached an
equilibrium frequency of about 0.1 in West Africa
because, although it killed the homozygotes, it
protected the heterozygotes against malaria. An
advantage of 11 percent of the heterozygote $\beta^A\beta^S$
over the $\beta^A\beta^A$ homozygote is sufficient to maintain
the gene β^S at about 10 percent frequency even
though the $\beta^S\beta^S$ homozygote has practically no reproductive fitness. In the black U.S. population
the β^S gene frequency is already much lower, because the survival disadvantage produced by hemoglobin S is no longer compensated by the selective
advantage of resistance against malaria, because

this disease is now almost absent from the United
States.

Geneticists use the concept of <u>fitness</u> to
denote the relative reproductive success, that is,
the relative number of offspring produced by a
given genotype. The effect of selection on a gene
can be defined more precisely in terms of a <u>selec-
tion coefficient</u> s, which is 1 when the homozygote
dies and 0 when it is fully normal. The value of
s determines the exit of a gene from the population. For a deleterious recessive gene with a
selection coefficient s, the gene frequency q decreases by $sq^2(1 - q)/(1 - sq^2)$ at each generation. As for the sources of entry of a given gene
into a population, they are two: matings with a
different population and mutation. The former
source only extends the range of spread of existing genes.

Selection is always at work, but the fate of a
gene depends not only on the external environment,
as in the case of the β^S gene, but also on the
genetic environment. Each gene is being tested
for fitness, that is, for the contribution it
makes to reproductive success, not by itself but
in a specific genetic background. To put this in
molecular terms, let us take a hypothetical case
about hemoglobin $\alpha_2\beta_2$. The four chains must
fit together in a special way in order to function
well. Suppose an α chain gene mutates: its
product α' may fit a bit less well with β. In a
population, however, there may be some polymorphism for β, and 1 percent of the population may
have an β' gene whose protein chain fits as well
or even better with α' than with α. Then, in
the course of random matings, some $\alpha'\beta'$ individuals are produced. If the hypothetical $\alpha'\beta'$ hemoglobin were a better hemoglobin, in the sense of
providing its possessors some advantage under some
circumstances, it might spread and become a major
hemoglobin type by natural selection.

Mutation is the ultimate source of all variation. Even though it is rare, 10^{-6} or less per
gene per generation, it supplies all new materials
for evolution. Assuming that most mutations produce genes that in the original genetic setting
are less good than the original ones, there will
be selection against the mutants. Under certain
conditions the rate of mutation and rate of loss
of the mutated gene will balance. This gives a

mutated gene a chance to be shuffled around many
times in mating and to be tried in a variety of
genotypes. In this way material for evolution
is preserved. It has been calculated that a human
being, with 20,000 or more genes, may be heterozygous for as many as 100 mutant genes that are
potentially deleterious in the homozygous state.
The evolutionary race is between the elimination
of these genes--by reproductive failure of homo-
zygotes--and their finding a setting where they
are useful.

Genetic drift

There is another class of events that can
alter the frequency of genes irrespective of mutation or of fitness. This is genetic drift, that
is, the chance that a gene or combination of
genes happens to be present in a small isolated
population at a frequency greater than its overall mean value. This founder effect magnifies
the differences in gene frequency. The following
is an interesting example. The desert island of
Tristan da Cunha in the South Atlantic was settled
in 1817 by a group of 15 people. When a volcano
erupted in 1961, all 260 people living there were
taken to England and were given medical tests.
Four of them turned out to be afflicted with
retinitis pigmentosa, a very rare disease of the
eye caused by a recessive gene. This was just
the frequency to be expected if one of the 15
founders were heterozygous for the gene and if at
each generation the marriages between the descendants had occurred at random. By chance, this
pathological gene was present 100 times more in
the small founder sample than in the average human
population.

For less deleterious genes, such as different
forms of hemoglobin, or PTC tasting, or allelic
forms of various enzymes, genetic drift due to
isolation of small groups of individuals can produce populations with very different gene frequencies. But drift is probably not the cause of
the great differences found among major populations of mankind, for example, the 12 percent
frequency of PTC nontasters in China versus 32
percent among Americans. A more likely cause of
such differences is the association of a gene,
the taster gene for example, in close linkage to
certain genes that became widespread in the
Chinese population because as a group they con-
ferred some selective advantage. One must always

keep in mind that, even though we can sometimes
define the selective coefficient for a gene,
especially for domestic animals or plants, natural
selection is working not on genes but on phenotypes. The fate of each gene depends on how it
functions in a variety of different genotypes to
produce phenotypes of different reproductive
efficiency.

Gene evolution

Genes have evolved as organisms have evolved.
One can trace, for example, the evolution of
various hemoglobin genes found in man from a study
of the amino acid differences in the polypeptide
chains.

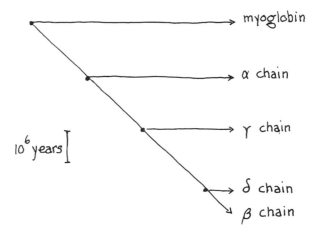

The interpretation is that a gene became duplicated
within a chromosome and that one of the copies,
hundreds of generations after the duplication,
underwent mutations and assumed new roles: myoglobin in muscles, α and γ genes in fetal hemoglobin, α and β in major adult hemoglobin, and α
and δ in a type of hemoglobin present as a small
fraction of the total human hemoglobin.

Gene duplication followed by mutation and ac-
quisition of new function by the corresponding
protein is the most commonly accepted explanation
for the origin of new genes. It can occur by a
mistake in the mechanism of DNA replication and
also by accidental illegitimate exchanges between
homologous chromosomes. There must have been a
great many such duplications in the course of
evolution to get to organisms like mammals that
have 1,000 times more DNA per cell than E. coli.
And many more, probably, to create the genome of
E. coli from primitive genes.

Some genes, however, seem to have remained
unique and quite stable over billions of years of
evolution. The best known example is the gene
that codes for the respiratory protein cytochrome
C. This is one of the iron proteins that ferries
electrons from NADH to O_2 (remember?). Biochemists have worked out the amino acid sequence of
cytochrome C, 104 amino acids, from 38 different
organisms, from yeast to plants to man. Some
parts of the molecule are identical or very similar in all organisms, other parts are more variable. Obviously, natural selection must have preserved those amino acids essential to maintain
the functional structure of the molecule,
especially the part that holds the heme group,
which is the electron-carrying part of the molecule. When "acceptable" substitutions occur,
they tend to replace an amino acid with a functionally similar one--aspartic acid for glutamic
acid, for example. If we consider the number and
location of differences between pairs of cytochromes, we can construct a map of relatedness.

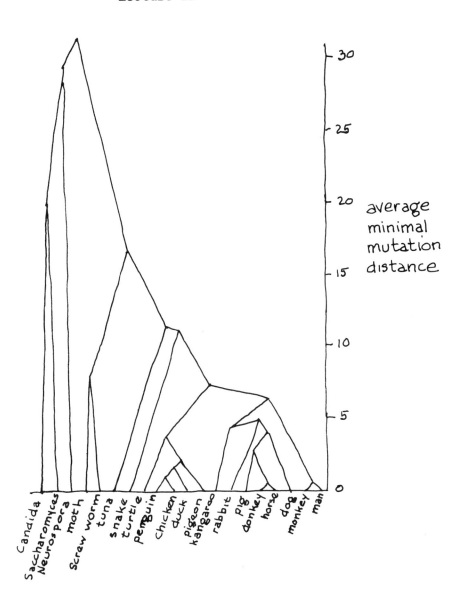

Speciation

It is rather pleasing that this "tree" fits the evolutionaly tree that paleontologists, zoologists, and botanists have arrived at.

As long as the members of a certain group of individuals can mate and produce fertile progeny, they are members of the same species. There is gene flow among them. If two populations of a given species become geographically separated, there occur progressive changes in allele frequencies at many gene loci because different alleles have different selection coefficients in different situations. Such isolated populations evolve into subspecies, between which the gene

flow becomes less and less. As more and more
mutant genes become established, the fertility
of matings between members of the two subspecies
decreases and so does the tendency to mate. The
original species has split into two new species.
This speciation by geographic isolation is by far
the most common mechanism for the origin of new
species of organisms. Much more rare, if it
occurs at all, is the origin of new species by
sympatric evolution, that is, by genetic diver-
gence within the same territorial area, for example, by persistent assortative matings.

At the other end of the story, there is the
fact that species die out. Those that exist
today are only an infinitesimal fraction of the
species that arose, prospered or just managed to
spread a bit, then petered out and disappeared.
Of these, we know only the ones for which we have
found fossil remains; but we may infer many others
as transitions between fossil forms and those that
exist now. An important point to remember is
that it is most unlikely that a distant ancestor
species persists at the same time as a descendant
species. The popular interpretation of Darwinian
evolution: "Man descends from the apes" is a
fallacy. Man and the higher apes now alive de-
scend from common ancestors whose time of existence
can be estimated at one or more millions of years.
Since then, several species of hominoids arose
and became extinct.

Contrary to the rather common belief that evolution has perfected every species for optimal performance under a permanent set of circumstances,
the modern theory of evolution, called neo-Dar-
winism, recognizes that evolution occurs because
natural selection is continuously--and often unsuc-
cessfully--battling to prevent extinction by increasing the proportion of individuals that perform
reproductionally better than others. The greatest
good for a species is to possess enough genetic
polymorphism to provide phenotypes that will have
reasonable degree of fitness in the varying envi-
ronments--including food, climate, competitors,
predators, infectious agents--that the species may
encounter. But in the long run each species that
exists is barely making the grade, barely surviving.
Evolution is the story of a few successes that have
emerged out of many failures.

Man has introduced something new in evolution,

both for himself and for other organisms, which it
domesticates or exterminates or preserves or up-
roots. One wonders whether, by using our brain-
power and our ability to alter the environment,
we shall be able to last longer than other dominant species have in the past. Certainly, there
has never been on earth any species so dominant
as man is now, not even the enormous but stupid
dinosaurs that roamed the earth 200 million years
ago. Man's chances of survival--for how long,
and in what numbers--are difficult to evaluate.
Evolution theory explains the past but is by no
means an adequate guide to predicting the future.
Too many uncertainties exist as to what conditions
will be 1,000 or 10,000 or one million years hence.

Further reading   Strickberger's Chapters 20, "Gene Frequencies
and Equilibrium," and 31, "Changes in Gene Fre-
quencies," cover many additional topics of population genetics.

PART IV

DEVELOPMENTAL BIOLOGY

Lecture 22

Development   At this point we must turn to embryology--how the
genetic program gives rise to the mature organism.
Some consider this the most exciting area of biology, others the major frustration; both are right.
Embryology is the study of development, that is,
of the unfolding of the program that exists in
the genetic material of a complex multicellular
organism. Embryology is exciting because the unfolding is an almost miraculous process. It is
frustrating because it is so difficult to interpret the miracle in precise biochemical terms.

Remember that a genetic program is not a plan.
This is seen clearly in development. It is nothing like an architect's design for building a
house; it is only a set of directions inscribed
in a tape, to be read at appropriate times in
response to appropriate signals, some coming from
the tape itself, some from the environment. The
alternative possibilities were classically considered under the dilemma of epigenesis versus
preformation: epigenesis, the sequential unfolding of a program; preformation, the actualization
of a plan. In its most extreme form preformation
postulated a miniature homunculus within the human
sperm (note the sexist implication).

The exciting aspect of development is obvious.
A fertilized egg generates a baby in 9 months, an
adult in 15 years. The 10^{13} cells of an adult
man have infinitely more variety than the 10^{13}
almost identical cells that one E. coli cell can
make in 24 hours. Each of my cells has its own
specific function. How does this come about?

Here the frustration comes in. We cannot give
an answer in terms of biochemical mechanisms. We
do not know what signals cause a liver cell to become and remain a liver cell. When we talked of
genetic diseases of man, we realized that their
essence was not only the genetic mechanism, but
also the developmental manifestation. Thus, the
study of development consists for the time being
in a description of processes, in a definition of
regularities, and in attempts to uncover underlying mechanisms.

Differentiation   After development, the second essential concept
in embryology is differentiation. A fertilized
human egg containing a diploid number of chromosomes divides repeatedly to produce 10^{13} cells.
Normally, all of the cells of an organism have
the same chromosomes. Hence differentiation to

form different types of cells is not due to a change in the amount of the genetic material contained in the cells. It must be due to differences in the function of the genetic material. Thus differential gene function is the basis of differentiation.

Gene constancy

How do we know that the cells in different parts of the body of an animal have the same amount of genetic material? One of the most significant experiments, done by Briggs and King, consisted of transplanting the nucleus from a differentiated cell of a frog tissue into a frog's egg whose nucleus had previously been destroyed by x rays or removed surgically. The enucleated egg cell that received a nucleus from a differen-

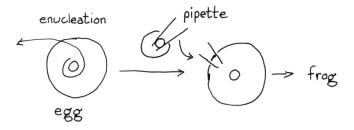

tiated frog cell could in at least a few instances develop into an adult frog. This means that the nucleus of the differentiated cell still had all of the genetic information needed to produce a frog. Nuclei from late embryonal or adult frog cells often gave abnormal tadpoles. But in a better experimental organism, the African toad Xenopus, even nuclei from adult cells when put into enucleated eggs can generate normal, sexually fertile adults in a reasonably high proportion of cases.

What about mammals and man? Experiments of the type done with frogs and toads have not yet been done. Mammalian eggs are much too small (about 10^{-1} mm in diameter instead of about 2 mm for the frog). They can be isolated from the ovary and have recently been fertilized with sperm in a test tube. No baby has yet been grown from such test-tubes zygotes by implantation into a woman's womb. In fact, the very suggestion that this might be done has roused justified concern about ethical problems in experimenting with human life. If nuclear transplantation into human eggs became feasible, even bigger problems would

arise. Would people want batches or "clones" of
N identical multiplets generated by transplantation of N nuclei from Mr. or Ms. X to N enucleated
eggs? I certainly would not like the idea of
such reproduction by cloning.

The evidence for genetic constancy in cells of
mammals is indirect. It comes from counts of
chromosomes, which usually remain in the precise
diploid number characteristic of the species, and
from measurements of amounts of DNA per cell,
which are precise to about 5 percent. There are
exceptions, however. One of them has to do with
the proliferation of special parts of DNA that
produce the ribosomal RNA needed to prepare an
egg cell for the early phases of development.
This specific gene amplification will be discussed
in one of the following lectures.

**Differential gene
action**

Apart from these special cases, all cells of an
organism have the same complement of DNA, of
chromosomes, and presumably of genes. Differentiation, therefore, must have to do with differential gene function. Can the mechanisms that
operate in bacteria explain differential gene
function in animal cells? In bacteria, individual
genes or groups of genes are turned on or off by
the action of specific proteins controlled by regulatory substances such as amino acid derivatives,
or sugars, etc. Such a mechanism can work well
in a single-celled organism like a bacterium because it provides a quick means of responding to
the environment. It depends, however, on the
fact that in bacteria the mRNA is unstable, de-
caying with a half-life of about 1.5 minutes.

But the messenger RNA in eucaryotic cells, from
yeast to man, turns out to be relatively stable.
It can persist for hours or days. For example,
red blood cells newly entered into the blood from
the bone marrow have no nucleus, but continue to
make hemoglobin because they contain messenger
RNA that decays slowly. What factors are responsible for the difference in messenger stability between procaryotic and eucaryotic cells is not
known. At any rate, the regulation of gene function must evidently be different, also because
its effects are often not reversible and because
the specific mechanisms involve turning on or off
many genes at a time. Out of the total set of
genes, there must be several subsets that function
specifically in different differentiated cells.

There must, of course, be substantial intersection
between subsets because many genes function in
more than one type of differentiated cell. There
is evidence that a class of basic proteins,
called histones, which are present in all chromosomes, play an important role in regulation; but
we know little of such role.

The range of gene expression in different
organs can be tested by hybridizing the RNA extracted from different types of differentiated
cells of an organism with the DNA extracted from
the cells of that organism and measuring the extent that the RNA from one organ competes with
that of another organ for DNA sequences. The
findings are as predictable: different types of
cells contain different classes of RNA, but certain RNA classes are found in many different
cells. Essential genes, like those involved in
producing components of the protein synthesizing
machinery, or of mitochondria, or of the cell membranes are presumably always turned on at least in
cells that grow and divide.

Determination

How does differentiation come about? At some
point in development, determination occurs. Determination is not the same as differentiation.
For example, certain masses of cells early in
embryonic development are already committed to a
later course of development. They are already
determined, although no recognizable differentiation has yet occurred. The specific fate of cells
of the early amphibian embryo can be followed by
using a vital dye (so called because it does not

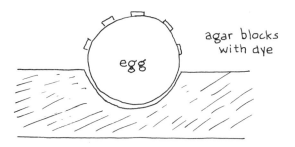

harm the cells) that can be absorbed in bits of
agar and applied on various positions of surface
of the embryo. The dye gets inside the cells and
marks them throughout development. The German
embryologist Vogt used this method to show that
each stained patch gave rise a certain part of the

embryo. Thus there was an early determination
in vivo of the destiny of these cells.

How fixed is the determination? Can the
destiny of a group of cells be changed? Sometimes yes and sometimes no. In vivo determination,
of course, is relatively stable, in the sense that
cells do not move around much. But the question
is whether determination is maintained only by
the proper positional relation of a group of cells
to neighboring cells or whether it is at some
point irreversibly fixed.

Traditionally the types of development in early
embryos were divided into two classes, mosaic and
regulative development. In mosaic eggs, development was supposed to take place so that the
destiny of a certain group of cells in a definite
location at a particular stage of development was
absolutely fixed. Each part of the mosaic egg
would represent a corresponding part of the adult
organism. Regulative eggs were those in which
different parts of the early embryo could change
destiny if their position was shifted.

The classic example of mosaic was supposedly
the frog embryo. Killing one of the first two
cells caused the other one to develop into a half
embryo. The classic example of regulation, interestingly enough, also came from splitting frog
embryo at the two cell stage: each of the separated cells gives rise to complete embryo. The
finding with the killed cell turned out to be due
to the presence of the dead cell.

The distinction between mosaic and regulation
is, in fact, not valid. We now believe that the
development of an embryo should be visualized as
a series of processes in which up to a certain
point the course of phenomena can be reversed
because the events are due to reversible interactions between cells, probably chemical and electronic exchanges. After a certain point in
development of specific parts of the embryo some
more stable changes occur, which affect cells in
an irreversible way and make their program more
or less irreversibly restricted. Reconversion is
no longer possible. In reality, both types of
phenomena occur in all types of eggs.

Morphogenesis

What makes a cell become different from another
cell? Cell contacts allow for chemical interactions between cells and may result in the turning on or off of certain groups of genes. Another

unit phenomenon is cell movement, the sliding of cells along one another. This plays an important role in animals but not in plants. Sliding of cells is part of <u>morphogenesis</u>, the creation of the form of the embryo. Differential growth and multiplication of different cells or groups of cells is important in morphogenesis. Finally, diffusible substances, starting from one point of origin, can by diffusion establish concentration gradients over 50 or 100 cell distances. Such gradients are presumed to play roles in determining what cells become in various parts of the gradient.

Sporulation

There are in organisms like bacteria some instructive phenomena that reveal mechanisms that can turn on or off large subsets of genes and possible play a role in developmental processes. Some bacteria are <u>spore formers</u>: when the food becomes depleted they engage into a slow process that leads to the formation of a spore inside the bacterial cell. The spore is a complicated entity containing many proteins that are not present in the growing bacterium plus one full set of the bacterial DNA. It is a resting cell, almost completely dehydrated. It can survive almost forever in a dormant state. When placed in the right situation, it converts itself again into a normal bacterium.

When a bacterium makes a spore, it must switch from making one subset of proteins to making another subset. The key to the switch seems to be a change that has been observed in the enzyme RNA polymerase, the transcribing enzyme. As a bac-

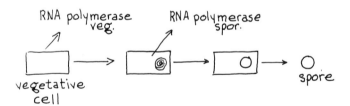

terial cell begins to sporulate, its RNA polymerase changes the efficiency of its transcription for different DNA templates--supposedly becoming more active on spore genes and inactive on many other genes. That the enzyme changes is reflected in its structure: of its usual five chains, 2α,

1β, 1β', 1σ, the β' and the σ appear to be altered.
The stimulus for these changes is clearly environmental, namely, the running out of certain foods.

It has been suggested (without any real evidence)
that a mechanism similar to that operating in
sporulation--a change in RNA polymerase--may play
a role in the development of higher organisms.
Perhaps, for example, slight changes in RNA polymerase could result in the transcription of different subsets of genes in different types of
differentiated cells. All of this would somehow
have to be regulated by the positioning of the
cells in the embryo.

Apart from the question of biochemical mechanism,
procaryotic organisms can tell us something about
factors involved in morphogenesis. Take the blue
green alga <u>Anabaena</u>, not very different from bacteria except that, as in most such algae, the
cells remain lined up together in slimy tubes.
Some cells become sporelike cysts; but not all do

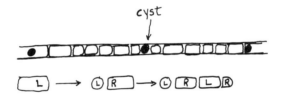

Rule: a cell that was left (L) at a division
gives rise to a small cell on the left
   (and vice versa for right).

so. Along the tubular filaments the cysts are
located at regular intervals. This means that
the organism has a way of measuring how far one
cyst is from the next, the presence of a cyst
preventing formation of another within a certain
distance. In this organism cell division is
asymmetric: each division produces one big and
one small cell. The small cell then grows in
size. At the next division, the new small cell
is always located away from the side of the previous division. Spores only form in small cells
before they grow.

There must be two signals involved: one is a
polarity signal, which tells each cell to divide
asymmetrically in a given direction; the other
signal, a gradient signal presumably coming from
the existing spores, prevents the formation of

other spores and keeps the cells growing and
dividing until the distance is large enough. Note
that cells must have receptors in order to respond
to these signals. We shall encounter other examples of such mechanisms in higher organisms.

One more case history. The slime mold
Dictyostelium (a eucaryotic organism) can construct from individual cells or amoebae a complex structure differentiated into a foot, a stem,
and a head, which contains reproductive cells.

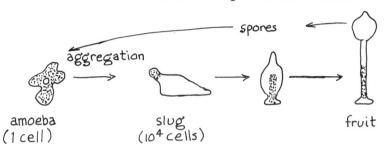

amoeba   slug   fruit
(1 cell)   (10^4 cells)

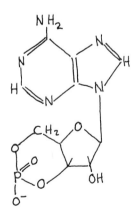

The normal structure is built by a specific
morphogenetic process in which about 10,000 cells,
after coming together into a mass called a slug,
take up specific positions by oriented movements.
There is a mutant of this mold whose slug has
only 10 or 12 cells, but these few arrange themselves to produce a perfect miniature structure.
Clearly, the intercellular interactions in the
mutant are reduced by a factor of 1,000, but the
pattern remains the same. Each cell must sense
the ratio of the number of cells above and below
it and behaves accordingly. Mechanisms like those
acting in Anabaena are probably at work here too,
but we do not know what they are. Slug formation,
by aggregation of cell groups grown on the surface of an agar plate, is a response to the concentration of a specific chemical, cyclic AMP,
that plays an important role in many regulatory
phenomena. This is an exciting finding: at least
one morphogenetic phenomenon, if only in a lowly
slime mold, can be attributed to a specific chemical. And, moreover, to a chemical that may have
continued to play such a role from the earliest
stages of life--from bacteria, in which it controls the regulation of the genes of sugar
metabolism, to human cells where it mediates,
among other processes, the hormonal control over
the release of sugar for muscle contraction. But

this will come later.

Further reading   Watson's Chapter 16, "Embryology at the Molecular Level," presents an unorthodox view of
developmental phenomena.

Lecture 23

Gametogenesis

The production of a diploid organism begins by generating the haploid gametes, egg and sperm, through the process of meiosis. Oogenesis is the name for the meiotic process that produces egg cells, spermatogenesis is the name for the process that produces sperms. In animals, eggs or sperms are produced from cells in specialized organs in the body; in plants, from cells that may arise in many parts of the body. All sorts of technical problems had to be solved by evolution in different organisms to make it possible for egg and sperm cells to come together. When they finally merge to form a zygote, the sperm contributes only its nucleus. The egg, which had passively awaited the sperm, contributes its nucleus plus practically all the cytoplasmic structures that will be used in the early stages of the embryonal development. This is the maternal phase of early embryonal life.

In man, sperms are formed in the testes by a regulated process. At birth, the testes contain cells called spermatogonia, the predecessors of the sperm cells. Until puberty, the spermatogonia divide extremely slowly and produce few or no sperms. Around the age of 11 or 12 years, under the influence of hormones, spermatogonia begin to divide rapidly producing millions of daughter cells. Some of these cells begin to undergo meiosis, each spermatogonium giving rise to four spermatids. The spermatids become sperms by a process of maturation involving the formation of

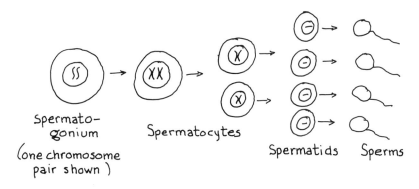

Spermato-
gonium   Spermatocytes
(one chromosome
 pair shown)   Spermatids   Sperms

a filamentous tail, which allows the sperm cells to swim, and the production of chemicals that make the sperms attracted to the female womb. Thus a sperm is a cell that has undergone a specific differentiation process, which converts it to a gamete ready to be transferred from the male to the female.

In the case of the egg the situation is a bit more complicated. When a woman (or practically any mammalian female) is born, all the eggs that she will ever possess are already present in the ovaries, as was mentioned before. The maturation from the oogonia to the oocytes (that is, the cells in prophase of the first meiotic division) is completed during fetal life. The oocytes stay put until puberty, when hormonal changes occur and the remarkable phenomenon of cyclic ovulation begins. In humans, once every 28 days (in other mammalians more or less cyclically according to different estrus controls) one egg (or sometimes two) matures: it pops out of the sac or follicle in which each egg is individually held in the ova ovary and starts descending the tubes toward the womb. If while going toward the womb the egg is

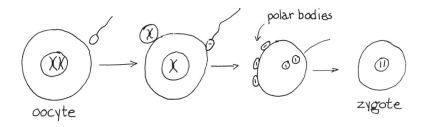

oocyte

polar bodies

zygote

fertilized, it completes the process of meiosis, but it throws away 3 of the 4 product nuclei as polar bodies. The fourth becomes the egg pro-nucleus. The sperm penetrates the egg cell, leav-ing its tail outside. The haploid male pronucleus and the haploid female pronucleus fuse and the di-ploid complement is restored. If an egg is not fertilized, it degenerates before completing the second meiotic division and dies. Therefore, the entry of the sperm activates the egg, preventing its degeneration. After fertilization and pro-nuclear fusion things start happening without delay. The nuclear membrane disappears and mito-sis occurs. The fertilized egg is on its way to developing into an embryo.

Plant embryogenesis   Let us now consider the situation in plants, specifically in flowering plants. Here things are different from what happens in animals and illustrate some important principles of differen-tiation of sexual organisms. Schematically, in flowering plants, the female meiosis leads to

formation of a sac that contains a haploid egg or
ovum with a haploid set of chromosomes. The sac
contains several other nuclei. One nucleus, in
particular, contains 2n chromosomes (diploid
number): It is derived from meiosis, but it
arises by the fusion of two haploid meiotic
nuclei.

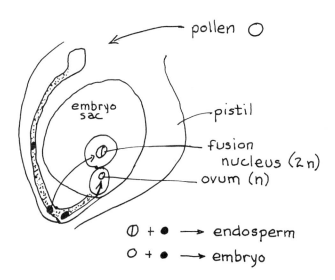

The male gamete, contained in a <u>pollen grain</u>
produced by male flower meiosis, lands on the end
of the <u>pistil</u> of a female flower and reaches the
egg cell through a process of germination in
which the pollen grain generates a <u>pollen tube</u>.
In different plants the tube may be from a few
millimeters to several feet long; in corn, for
example, the silk of the ear represents the pis-
tils through which the pollen tubes must grow to
the eggs in the ear itself. The pollen tube contains just three nuclei. Of these, one nucleus
disintegrates (function unknown); one fuses with
the oocyte to form the zygote, which then develops into the embryo; the third fuses with the
2n nucleus in the egg sac to form a triploid cell
that gives rise to the <u>endosperm</u>. The triploid
endosperm has no counterpart in animals. It is
the flesh of the apple or the starchy part of corn
or wheat seeds. In a coconut, the outer part of
the endosperm is the white rind we use shredded
in coconut cookies; the central part liquefies
and becomes coconut milk, which contains plenty
of plant growth factors and some suspended cells.
Coconut milk is a favorite of plant physiologists

as a good natural medium for plant tissue culture.

Thus the sexual process in flowering plants generates a dual system: one is the embryo, the other is a tissue, the endosperm, designed to feed the embryo in its early development, which may last for weeks, months, or years. Once a plant has grown from the embryo, the endosperm is no longer needed, because the plant can obtain nutrients from photosynthesis and from its roots.

The arrangement in the flowering plants differs from those found in animals in that the tissue that provides food to the growing plant, the endosperm, is generated by a special by-path in fertilization. Animal embryos receive their early food from the yolk which is part of the egg cell.

Some plants have arrangements other than the one just described: they get their earliest food from the cotyledons, which are derived from the zygote rather than the endosperm nucleus.

The cell divisions that produce the plant organism do not require cell movement. This is an important difference from animals, in which, as we shall see, the construction of the embryo requires not only cell division but also cell movements, displacements, and sliding. In plants all differentiation originates by differential reproduction. For example, the first step in the embryonal development of flowering plants is an asymmetric division: one of the daughter cells gives rise to the plant; the other produces the suspensor, which holds the embryo attached to the wall of the embryo sac (this later becomes the wall of the seed). The daughter cell that generates the plant again produces an asymmetric structure consisting of three layers of cells, the procambium, the cortex, and the epidermis.

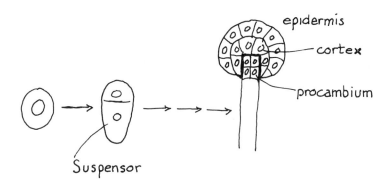

What causes the differentiation of these early
embryonal cell types, whether hormones from
various parts of the seed or gradients of substances produced from one or more sites in the
seed, we just don't know.

When a seed with its embryo finally germinates,
all growth of the embryo to form the plant occurs
at the root and apex. This brings out another
fundamental difference between plant and animal
development. A plant potentially can grow forever. It is programmed that way because from the
very earliest stages it delegates the ability to
undergo mitosis to a certain set of nondifferen-
tiated cells called meristems. There is an apical
meristem, a root tip meristem, and later, when
shoots begin to form, additional meristems in
each shoot. In the stems of some plants meristematic cells are located in lateral layers so that
growth can result in an increased diameter of the
stem. The trunk of a tree, seen in a very schema-
tic way, contains a central core of wood derived
from the xylem, consisting of tube cells which are

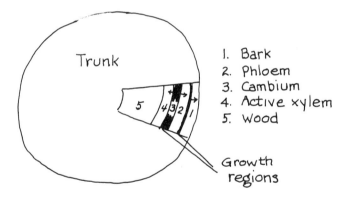

1. Bark
2. Phloem
3. Cambium
4. Active xylem
5. Wood

involved in transporting water from the roots and
which soon die and become part of the wooden core.
Next to the xylem is a layer called the phloem,
through which water loaded with food produced by
photosynthesis circulates up and down the plant.
Finally, the trunk is surrounded by bark. Between xylem and phloem there is one layer of meristematic tissue or cambium, which produces xylem
cells toward the inside and phloem cells toward
the outside. A second layer of meristematic tissue lies just under the bark and produces new
bark cells. Note that the stem structure effects
the production of different cell types from the

two sides of the lateral meristem. This implies
that different controls, hormonal or other, must
operate on the two sides.

In essence, then, the structure of most plants
is rather simple. Throughout the life of the
plant, a reservoir of embryonal cells is maintained. Annual plants die not because of loss of
meristematic tissue but because the atmospheric
conditions become intolerable to them. Any meristem cell that is not dead is potentially able to
develop into an entire new organism. For example,
a plant can be started (note the difference from
starting a baby!) from a cutting placed in water
provided the cutting contains at least one active
shoot. Better still is to use water plus rooting
substances, which are really plant hormones. Once
roots form, the new plant can then be placed in
soil.

Plant segments, therefore, have autonomy
thanks to the presence of the meristematic tissue.
Disassociated cells from meristematic tissue or
from embryos can be cultured in coconut milk, and
individual cells can then form embryos, which if
handled carefully can grow into complete plants!

Growth limits

A beech tree or an oak can grow to enormous
sizes. What keeps humans from growing into giants?
How is the normal size of an animal species defined? The answer must be related to the hormonal
explosions that occur toward the end of the growth
period, usually at the time of puberty. Similar
hormonal explosions are not observed in plants
except at the time when the plants come out of the
seed. Otherwise, the hormonal situation in plants
appears to be always a local one. Note that many
animals, like crustacea and some fish, continue
to grow in size throughout life if enough food
and time are available.

In flowering plants the flowers are found all
over the plant when hormonal and climatic stimuli
occur. There is no such thing as germ cells
being set aside early in development to give rise
later to gametes, as there is in animals. Regulation controls growth and flowering. The top of
the plant sends inhibitory stimuli so that other
parts grow less. In my backyard the top shoot
of a young spruce was growing crookedly. I broke
it off and one week later one of the lateral
shoots had turned vertically and taken over as
the apical shoot.

A plant may be able to produce only male, or only female gametes, or both types. Ginko trees, for example, are male or female, not both. At the University of Illinois we had a female tree next to the Chemistry Building and some male trees at the other end of campus. When there was enough wind to bring pollen across campus, ginko fruits developed. When they ripen, these fruits contain an enormous amount of butyric acid, and the stench in the fall was terrible.

Bisexual plants may or may not produce bisexual flowers: male and female flowers may occur in different places on the same plant. In corn, for example, the tassle on the top of the plant is the male flower and the ear is the female flower. The silk of the ear receives the pollen, which may have to produce tubes a foot long.

Even plants with bisexual flowers often have devices to prevent or reduce self-fertilization. Plants seem to have reached the same conclusion as geneticists did: self-fertilization is undesirable. They have many physical or chemical ways of preventing it: the pistil may be so far out that the pollen cannot fall on it unless transported by a bee to another plant; or the pollen may mature at a different time than the seed; or growth of the pollen tube to the egg sac of the same flower may be prevented by some genetic self-sterility mechanism. Suppose self-fertilization occurred without interference (as in fact it does in beans and peas). Would all the progeny of a plant be identical? What would be the genetic consequences? Assortment of genes would occur during meiosis and some progeny of different types would be produced. Progeny plants carrying in homozygous form some deleterious genes would be eliminated. If self-fertilization continued over and over the progeny would tend to become more and more uniform. Such a lineage of identical plants would offer less range of variability to adapt to new climatic or nutritional conditions. Inbreeding is always potentially dangerous to species under natural conditions, no matter how effective it may be in producing new varieties of domestic plants or animals in order to satisfy specific human requirements.

Further reading   Ebert and Sussex' Chapter 5, "The Plant Embryo
                         and its Environment," is an elementary but detailed
                         discussion of plant embryology.

Lecture 24

Animal development

We now turn to animal development: how to make a fruit fly or a frog or a human being from an egg plus a sperm. What we want to do first is to identify, on the basis of various experiments, the specific contributions of the two gametes and their parts to the early stages of development. What is the role of the nucleus? What do various parts of the cytoplasm contribute, and when do these contributions cease? What is the relation between cell division and differentiation? We shall have to get our evidence where we can find it, using a medley of organisms and of experiments.

Let us start with the unfertilized egg. It is one of the largest cells in the body, even in mammals where it is only about 0.1 mm in diameter. Insect and amphibian eggs can be several milli-meters in linear dimensions; bird's eggs are much larger. Why the difference? As you already know, the large eggs have food stored as yolk and other materials, which supply the embryo during exten-sive stages of development. When a chick emerges it is an elaborately developed animal and so is a tadpole, although smaller. The mammalian egg needs much less yolk because, after it is ferti-lized, it soon starts receiving food from the mother's womb. In animals as in plants there are successive phases of development, which really represent different organisms, both in shape and in function, like the tadpole and the adult frog or the insect larva and the adult. These phases succeed each other in development and have their own food sources. A mammal obtains food first from its little bit of yolk, then from the mater-nal supply of blood, then from its own eating.

Parthenogenesis

What about egg components other than yolk? Let us take the nucleus first. The haploid female pronucleus fuses with the male pronucleus to give the diploid zygote nucleus. Is this essential? Not if all we want is to start an egg going. A frog's egg pricked with a needle will start to develop, and even a rabbit egg in the test tube can be made to divide by heating. This is called parthenogenesis and can lead to the formation of mature animals. The chromosome number, however, again becomes 2n because at one of the first mitoses, either the second meiotic division or the next one after that, the chromosomes fail to separate. We get therefore a fully maternal animal. [Can you see why all parthenogenetic rabbits are females?]

Parthenogenesis is sometimes a normal feature of a group of animals. In wasps, fertilized eggs give rise to diploid animals that are females, whereas unfertilized eggs produce haploid males. Usually, an egg needs activation by a sperm or other trigger; in wasps, however, activation is caused by the act of laying the eggs. [Notice that not all the haploid male wasps need be genetically identical; but the haploid condition causes exposure of all recessive genes that are present and therefore these males are subject to all sorts of lethality and unfavorable selection.]

Role of the nucleus

The most telling experiments about specific roles of the nucleus in development come from extremely different sources: the unicellular alga Acetabularia and the newt egg. Each Acetabularia cell has a root base, a stem and a cap. The nucleus is in the base. If one cuts off the cap, the stalk regenerates a new cap after a few months. (The cap does not produce a new stalk or root.) If both cap and root base are removed, the stalk slowly makes a new imperfect cap. What does the nucleus provide, then? We get the answer by cutting and cross grafting Acetabularia of different species that make different shaped caps. The first cap that is regenerated is intermediate between the two types; if

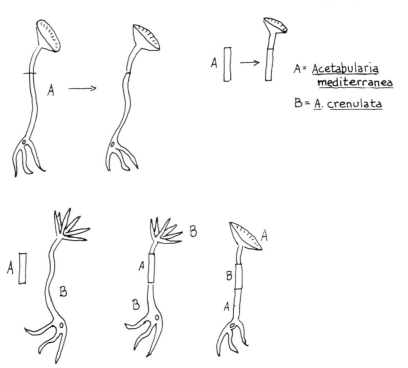

A = Acetabularia
      mediterranea

B = A. crenulata

this cap is removed, the next cap is always
characteristic of the species that provides the
nucleus. The nucleus evidently makes the morphogenetically specific substances. Apparently the
stalk contains enough morphogenetic products from
its former nucleus to sustain about one cap regeneration by itself. What are these products?
We do not know.

The evidence about the role of the nucleus in
development derived from the amphibian eggs comes
from some famous experiments by Spemann and his
students. In one case, Spemann tied a knot
around a newly fertilized newt egg with a loop of
hair, leaving a bridge of cytoplasm. The side of
the egg that contained the nucleus developed into
a ball of cells, which ultimately became a whole

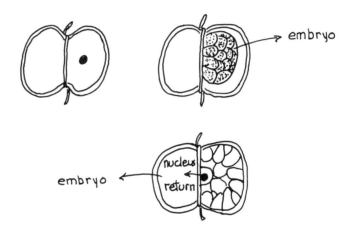

embryo. The other half never divided--unless,
after the nuclei had become smaller at the 4th or
5th division, one of them slipped through the
cytoplasmic channel into the other half egg. Then
this half egg started dividing and became an embryo, although delayed in its development. This
tells us, first, that the nucleus is needed for
division and embryo formation; second, that a
relatively late nucleus can support a normal
development. We already have learned this from
the experiments on transplantation of nuclei from
various body cells to enucleated eggs of frog or
newt.

Cytoplasmic differentiation

It is not surprising that the nucleus should
provide essential information for development.
After all, it has all the genes. A more interesting question is what the egg cytoplasm does. Even

more subtle is the question of whether specific
regions of the cytoplasm have information for the
development of the organism and whether one can
relate specific parts of the egg to specific
parts or organs of the adult. We shall later
come upon several informative pieces of evidence
in which positional information within organisms
seems to determine the course of development.
Here we are concerned specifically with positional
phenomena in the egg cytoplasm.

Let us take first the sea urchin egg. This
echinoderm develops first into a larva called
pluteus, with bilateral symmetry (the "first
organism," like the larva of an insect), then into a radially symmetrical adult. Sperm and eggs
are released in the water, where they meet. The
egg has a band of pigment that defines two poles,

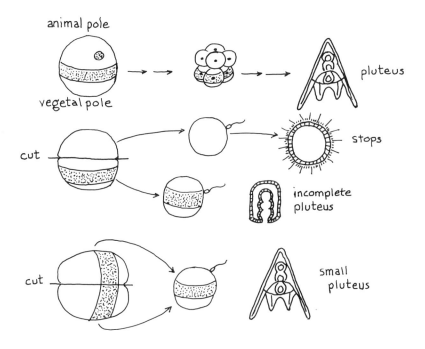

the so-called vegetal and animal poles. If you
cut a sea urchin egg in two, the two halves close
up and survive. One half has the female nucleus,
the other does not; but this is not really important. If a sperm fertilizes one of the two
halves, the result depends on whether the cut was
above the pigment band or through the band. If

the cut goes through the band, so that both
halves receive some part of the band, fertilization of either half generates an embryo that
develops into a pluteus (half of these plutei are
haploid, half diploid, of course). If the cut
was above the pigment band, after fertilization
the lower, vegetal half develops into an abnormal
pluteus; the upper, animal half becomes an abnormal vesicle or blastula and then stops.

Thus, irrespective of nuclei, some parts of the
egg cell contain substances needed for the proper
development of the embryo, and these substances
are not free to circulate inside the egg. They
may even be thought to reside stably on the
membrane of the egg.

The influence of localization of cytoplasmic
substances on development need not be limited to
localization preexisting in the egg. In amphibian
eggs there appears after fertilization a band, the
gray crescent, located opposite the point of entry
of the sperm. As we shall see later, the location
of this band plays a critical role in localizing
embryonal organs. In this instance, therefore,
the critical cytoplasmic localizations arise not
independently of fertilization but in response to
it. Again, we know nothing of the chemistry involved.

Insect development   Now for information from insects, especially
Drosophila. The fertilized egg generates a larva,
which then becomes a pupa or cocoon inside which
the adult, called the imago, is produced. The
body of the imago is formed by the imaginal disks,
groups of cells that are set aside early in embryonal development and then stay put without any
function throughout the larval stage [the biological equivalent of a bank trust that you cannot
touch till you are 21 years old]. When the larva
encases itself in the cocoon, the imaginal disks
start differentiating, each of them giving rise to
a specific part of the adult body. The larval
structures are dissolved and their materials are
used as food.

The relevant point is that the cells that are
going to produce the larval body, as well as those
that form the imaginal disks, become determined
very early within the egg itself. At the start,
as usual, the fertilized insect egg has two
haploid nuclei, one from the male and one from the
female. These nuclei fuse and a series of eight

or nine nuclear mitoses follows without cytoplasmic divisions, yielding a <u>syncytium</u> of about 500 nuclei all in the center of the egg cell. Within the syncytium, the nuclei do not mix or turn around, but continue to be oriented according to a constant vector, whose direction is dictated by the random orientation of the first mitotic spindle.

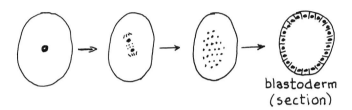

blastoderm
(section)

After the first 500 nuclei have been produced, they all move to the periphery of the egg, where they divide once or twice more, still forming a single layer. At this point each nucleus becomes a separate cell: it acquires a cytoplasm and surrounds itself with a cell membrane. The single layer of cells is called the <u>blastoderm</u>. The cells of the blastoderm give rise to all the organs of the larva and of the imago. The cells that will eventually give rise to a specific imaginal disk, for example, are already specified at this stage. If one destroys a few cells with a hot needle the corresponding adult organ will be missing.

The question is, What specifies the fate of a given cell? Is it something that occurs in the nuclei as they divide in the syncytium or as they reach the cell surface to form the blastoderm? The answer is that the specification depends only on where a given nucleus becomes located on the egg surface, which in turn depends only on the orientation of the first mitotic spindle. It is the cell surface that is preprogrammed to impress upon each nucleus its future destiny.

Gynandromorphs

The evidence for this goes back to findings by a geneticist named Sturtevant. He was interested in what happens if during development some cells of a <u>Drosophila</u> female lose one of their X chromosomes and therefore become male; that is, when flies become mosaic <u>gynandromorphs</u> (part female, part male). Sturtevant bred females that had one normal X-chromosome and one circular $\overset{\circ}{X}$ such as occasionally is produced after exposure to x rays.

Half of the eggs of such females have the ring X chromosome. When these eggs are fertilized, there

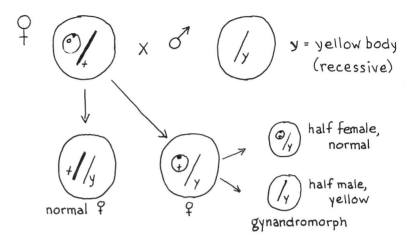

gynandromorph

is a substantial chance that the circular chromosome will get lost from one of the daughter cells at the first mitosis. When this happens, one of the daughter nuclei has two X's and is female ($\overset{\circ}{\text{X}}$X), the other has only one X (derived from the male parent) and is male (XO). This generates a gynandromorph with a half-female and a half-male body; the different halves can be recognized if they exhibit different genetic characters. For example, a recessive gene y (yellow body) in the X chromosome derived from the father is expressed on the male half XO and is masked by the dominant allele y^+ in the female half $\overset{\circ}{\text{X}}$X.

In individual gynandromorphs the plane that separates the male half from the female half turns out to be oriented in any direction, the direction being determined only by the orientation of the first mitotic spindle, which provides the vector for the plane of separation between male and female parts. Since there is no mixing of nuclei in the syncytium, the male and female parts remain coherent.

Apparently the 500 nuclei that are made in the syncytium and migrate to the egg surface were completely indifferent as far as their ultimate fate is concerned. The place on the egg surface where a nucleus lands, alone determines its destiny. There must be in the surface of the egg a <u>prepattern</u>, which determines what the individual cells in the blastoderm are going to become. The prepattern may exist before the nuclei migrate to

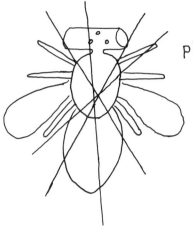

plane of male/female
separation random,
related to axis of
first nuclear cleavage

the surface. It may have been impressed in the
ovary by the cells that surrounded the egg, or it
may be established later. The essential thing is
that the entire surface is prepatterned, possibly
by gradients of substances coming from specific
sites of the egg. The information for differentiation is in the egg cytoplasm, specifically in
the surface layers of the cytoplasm.

Gynandromorphs even make it possible to decide
where in the fly's body a gene must function in
order to express its character in the adult fly.
For example, a gene X may affect wing operation
either by acting in the brain or by acting in the
development of the wings. A gynandromorph with
the N, normal allele in the ring chromosome and
the n allele in the paternal X chromosome flies
normally whenever the border between male and
female parts allows the N allele, that is, the
female genotype, to be present in the region of
the body where it is needed. One can use this
system to prepare a gene-function map of the
blastoderm, that is, a map showing where in the
course of development a given gene must act for
its function to be expressed. Unfortunately,
this cannot yet be done with mammals.

Germinal particles   Apart from the question of localized signals
for long-term differentiation in the egg cytoplasm,
we may ask questions as to the short-term role of
specific cytoplasmic structures in the early development of the fertilized egg. In the egg of
the toad Xenopus there are microscopically recognizable particles whose function has to do with
determining the future germ cells. As the fertilized egg divides, the particles remain concen-

trated in a few cells at the bottom. Later the
cells that have the particles migrate to a special
position inside the embryo and give rise to primary germ cells. If one destroys the particles,
for example by exposing the bottom of the egg to
ultraviolet light, the sex organs do not develop
at all. Because they are sensitive to ultraviolet light, the particles are supposed to contain RNA but even this is not at all certain.

Gene amplification

A better understood egg cytoplasm feature in
Xenopus is amplification. It has to do with a
mechanism that provides the egg cell with the
large amounts of ribosomes needed to make all the
proteins that the egg must produce before the
genome of the embryo itself is capable of taking
over the control of ribosomal synthesis.

Ribosomal RNA (rRNA) is made in the nucleolus
by a group of chromosomal genes to which the
nucleolus is attached. In the cells of Xenopus
as in those of most organisms there are several
genes for rRNA arranged in tandem, about 500 per
set of chromosomes. The RNA molecules they make
are of relatively large size, but are then split
into 2 main pieces, one used for the larger subunit of the ribosome, the other for the smaller
subunit. By electron microscopy one can see that
these genes are located in tandem on the chromosomes, separated by nontranscribed "spacer" regions, which may be the recognition sites for
RNA polymerase.

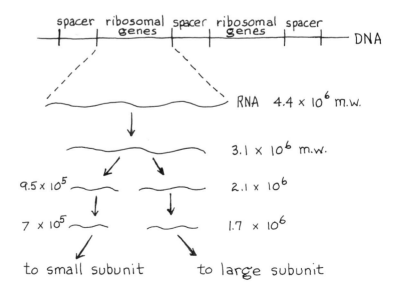

Normally, all cells have one or two nucleoli. The oocytes of Xenopus are special. They contain not one or two nucleoli, but thousands of them. These are loose in the nucleus and contain not only RNA but also DNA. The presence and nature of this DNA can be studied in a variety of ways, but the most useful is hybridization with labeled ribosomal RNA: that is, one extracts the DNA, separates its strands, mixes them with labeled rRNA, and measures how much rRNA can be complexed by a given amount of DNA. Since the RNA from ribosomes consists of rRNA only, the method can measure the amount of ribosomal genes. In somatic cells of Xenopus the genes for ribosomal RNA represent about 1 part in 1,000 of the total DNA.

It turns out that the oocytes of Xenopus have an enormous quantity of DNA specific for rRNA: about 31 picograms compared with 12 picograms for the entire DNA of one gene set. There has been a 3,000-fold amplification of this specific class of genes in preparation for embryonal development. One can verify that this DNA is actually rRNA-homologous DNA in the nucleoli by in situ hybridization. Slides with cells in agar are treated with alkali to separate the DNA strands and then radioactive rRNA is applied: it sticks to the nucleoli only and not to other parts of the cells. Autoradiography shows thousands of silver grains over the nucleus of mature oocytes.

somatic   early   mature
 cells   oocyte   oocyte

How is the amplification process regulated? We do not know. Nor do we know yet why amplification occurs in oocytes and not at other stages of development. Amplification has been looked for in cells other than oocytes, for example in the precursor cells of red blood cells, which make enormous amounts of hemoglobin and in the cells of the silkworm that make tremendous amounts of silk protein. None has been found.

Amplification occurs in the oocytes of

amphibians, fish, insects and clams. In Xenopus, the extra DNA disappears either before or during fertilization. In Drosophila and other insects, amplification occurs by a different mechanism, like a bridal shower: a group of nurse cells, which surround the egg, become polyploid and very large in size and transfer many of their ribosomes to the oocyte. Evidently, the requirement for piling up of ribosomes in the egg cells has existed in all organisms and has been solved, as usual in evolution, by different adaptive devices.

Further reading

For this lecture and Lectures 25, 26, 27, 28, the Ebert and Sussex book provides an elementary but stimulating supplement.

Lecture 25

Maternal effects   In the last lecture we saw how the egg cell prepares for the early synthetic needs of the embryo
that arise if the egg is fertilized. In addition
to extra synthetic machinery, the oocyte already
contains lots of messenger RNA made on the loops
of the chromosomes. This maternal messenger is
used for protein synthesis in early development;
the fertilized egg, in fact, can develop for a
while even if new RNA synthesis is blocked at the
time of fertilization by the addition of the substance actinomycin D, which inhibits all RNA synthesis. There is a complication, however. The
maternal RNA comes from the maternal genes, which
are not homozygous with the genes of the fertilized
egg. Hence there may be <u>maternal effects</u> on the
phenotype. These are usually expressed only in
the first few hours or days of development, but
sometimes they influence the entire life of the
offspring. A good example is that of the pigment
gene A in the moth <u>Ephestia</u>. The cross of a female
aa by a male aA gives the normal 50-50 ratios of
dark-eyed (A) to red-eyed (a) offspring. The
cross of a female Aa by a male aa, however, gives
all dark-eyed animals. The reason for this is
that the maternal gene A, functioning during the
oocyte's development, produces enough pigment
precursor to supply the progeny flies, including
those that came from an a oocyte and an a sperm.
[Mother, mother: can't you even let me make my
phenotype by myself?]

   In other cases the oocyte influences the phenotype of the new organism not by a lasting chemical
influence but by a vectorial influence in the
early development. A famous instance is that of
the snail <u>Limnaea</u>. This animal has shells that
are either right-handed or left-handed spirals.
The right-spiral gene is dominant, the left-spiral,
recessive. But the orientation is always dictated
by the maternal genotype: only the progeny of
left-left snails (homozygous recessive) has left-
spiral shells. The reason is that in these snails
the early cellular divisions after fertilization
are arranged spirally within the egg, and the
direction of the spiral is dictated entirely by
the egg cytoplasm. There are no phenomena of this
kind known in mammals yet, but it would be surprising if some did not exist.

Cellular
differentiation   We may now approach the problem of the elementary phenomena of development and differentiation

in another way, not at the level of the components of the fertilized egg but of its descendant cells. What are the relative potencies of various cells in the embryonal and adult organism, given that their nuclei are in first approximation equivalent? That is, which cells may still generate many different cell types and which are irreversibly committed to a given destiny?

We already know that in sea urchins and in amphibia one of the first two (or four or eight) cells are totipotents, that is, they can produce entire if smaller animals. For mammals like man, nature has done the experiment by producing identical twins, which come from a very early splitting of a fertilized egg. At the other end of the story, in the adult body we have cells with variable degrees of reproductive and differentiative potency. Take for example three organs, the brain, the liver, and the skin. Brain nerve cells, once mature, are permanently incapable of dividing, even when placed in culture. Liver cells normally do not divide after the body reaches its full normal size. But if you cut off a piece of rat liver, even 90 percent of it, most of the cells that remain enter the reproductive cycle, divide, and in a few days have restored the normal size of the liver. These cells, therefore, have retained reproductive capacity, which has been kept under control by hormonal signals from other liver cells. Also, the liver cells cannot differentiate; when released from inhibition they give only liver cells.

Stem cells
    The skin situation is different. There are in the basal layer of the epidermis individual stem cells, which divide all the time and generate a new stem cell and a differentiated cell. The latter, when it divides, gives rise to cells that cannot divide any more, move away from the basal layer, become horny by making the protein keratin, and finally die and scale off. The stem cells, although they reproduce all the time, are committed to produce only skin cells. They are programmed irreversibly. As we shall see, the skin programming is done early in development by the juxtaposition of certain groups of surface cells with special areas of connective tissue. (Does this remind you of the meristems in plant development?)

The stem cell mechanism is quite common in the organism. The potency of stem cells, however, is

not always restricted to the one single path of
differentiation. A very important case is that
of the hematopoietic system, the system that produces the cells of the blood, the bone marrow,
and the lymphoid organs such as the spleen and
lymph nodes. Early in the embryonal development
some groups of cells are set aside from the wall
of the primitive intestine to become the precursors of the hematopoietic system of the adult.
These cells then reproduce without becoming further differentiated, and even in the adult some
of their descendants remain as undifferentiated
stem cells in the bone marrow, either in the
resting state or dividing. But continuously,
from early fetal life to adult life, some stem
cell descendants are becoming committed to various pathways of differentiation to produce red
cells or any one of the various types of white
cells. A committed cell can still divide, but
its descendants are all the same kind of cells.

C = commitment

For example, the stem cells that are committed to
give rise to red cells do not actually do so until
they meet a substance called erythropoietin. The
level of erythropoietin determines the quantity
of red cells produced from those stem cells that
have already become committed to red cell production. How can one be sure that some stem cells
were already so committed? Because the stem cells
can be made to grow into isolated colonies; many
of these colonies contain only red cell precursors; some others contain only white cell precursors, and some--derived from uncommitted stem
cells--can still produce both types of cells.
    The point of all this is that, at various points
in development, there occur certain events that
convert some totipotent cells of an undifferentiated type to cell lines with more restricted
potency. Within each such line some descendants,

in response to local or humoral stimuli, become
committed to narrower and narrower fates. Finally
one gets to mature cells, which either have lost
even the ability to divide, like red blood cells
or nerve cells, or retain it but do not normally
use it, like liver cells or some white blood
cells. The inability of certain mature cells to
produce DNA and RNA--for example, the hen red
blood cells, which have synthetically inactive
nuclei--is not completely irreversible. This can
be shown by cell fusion, a technique by which cells
in culture can be made to fuse together by the
action of a certain virus. When hen red cells
fuse with human or hen or mouse cells, their nuclei
start functioning again: autoradiography shows
that they start making DNA and RNA. Some substance
derived from the cytoplasm or the nucleus of the
other cell must have shaken the red cell nuclei
out of their lethargy.

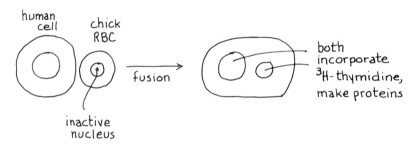

It is important to remember that the progressive
commitment of stem cell types goes on throughout
the embryonal development. This must represent a
permanent limitation of the set of genes that will
remain functional. Until we understand the meaning and the causes of this commitment, a complete
description of differentiation in biochemical
terms will not be feasible. That is why the study
of stem cells and their differentiation is at the
center of developmental biology. Cancer cells,
for example, are best seen as resulting from errors
of commitment.

Teratoma

Is it possible to isolate, culture, and study a
line of cells that are and remain totipotent--what
one might call a cancer of the zygote? Nature in
collaboration with mouse geneticists has provided
an answer. There is a class of tumors, called
teratomas and found in the testicles of certain
mammals, whose component cells represent a variety

of types resembling skin, intestine, nerve tissue,
etc. When one puts isolated cells of teratomas
into culture, some cells turn out to be totipotent:
that is, when reimplanted into a mouse they generate tumors in which all sorts of tissues differentiate, corresponding to the various types to which
a zygote gives origin in the course of embryonal
development. These totipotent teratoma cells continue to multiply in culture without differentiating; only when injected back into mice do they
undergo commitment and differentiation. Moreover,
the totipotent cells show on their surface some
specific substances that are present only on the
sperm cells and on the first few cells of the
embryo before it starts differentiating.

This system may offer a wonderful opportunity
to study early differentiation and to answer
experimentally a whole series of basic questions:
What substances cause a teratoma cell in the body
to become committed in one or another direction?
Where do these substances come from? What does
the response of the cells to these substances en-
tail? What genes are turned on or off?

The T locus   There are some known facts that throw some
light on the relation between cellular commitment
and differentiation on the one hand and gene expression on the other. One interesting set of
data relates to the teratoma story. Mice have a
group of 12 closely linked genes called the T
genes. For a mouse to be normal, all these genes
must be either homozygous or heterozygous for the
wild-type allele.

The T genes produce substances that are present
on sperm cells, on early embryonal cells, and on
teratoma cells. If any one of the 12 genes is
homozygous recessive the mouse's development is

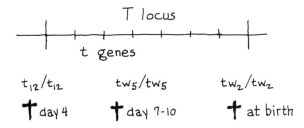

abnormal. For each of the 12 genes, homozygosity
of the mutant allele causes the embryonal development to stop at a precisely specific stage. The

earliest gene causes development to stop after
the first few cell divisions, before the first
ball of cells, called the morula, turns into the
vesicle called the blastula (which will be discussed later). One gene causes death at the time
of birth. Apparently the normal function of each
of these genes is needed and needed only at a
specific stage in early development. They are, in
a sense, gate-keeping genes for the sequential
process of embryonal development. The nature of
their products and the targets of their actions
remain to be discovered. There are probably many
other genes that function in a similar way.

Puff activation

There is an unrelated line of experimentation
that links sequential expression of genes with
sequences of developmental events. This one comes
from insects, specifically from Drosophila. In
the larva of this fly there are large salivary
glands, whose large cells have in their nuclei the
so-called giant chromosomes. (They are giant because they are <u>polytenic</u>, that is, they have many
copies of the chromosomal filament bundled side-by-
side.) These chromosomes are visibly banded and
each band is identifiable by its morphology.
Their sequence is recognizably congruent with the
positions of specific genes in the chromosome map;
each band corresponds to one or a few genes,
separated by an interband whose DNA may have only
regulatory roles. Occasionally one sees a band
swollen into a <u>puff</u>, a structure that contains
lots of RNA and is supposedly the site of a gene
that is actively functioning.

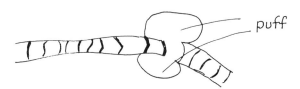

At each given stage of the fly's development the
pattern of puffs is constant: that is, puffs appear and then disappear, just as we expect some
genes to function at certain stages, others at
other stages.

A clever experiment (by a scientist named
Clever) consisted in injecting the hormone ecdysone, which is the major regulator of insect
development. Ecdysone caused two puffs to appear
within one hour and some others to come up 8 to 12

hours later. If, after adding ecdysone, he added an inhibitor of protein synthesis, Clever found that the early puffs did appear, but the late ones never formed. The interpretation is that ecdysone acts more or less directly in activating the genes corresponding to the first set of puffs. Then the products of one or more of the early activated genes in turn activate later genes. That is, genes act in cascade fashion, in a way that could be responsible for an orderly sequence of developmental stages.

The substance of this lecture can now be summarized. The zygote formed by fertilization generates cells that are totipotent but whose descendants at various times become committed to narrower potencies. This commitment generates differentiation and must derive from a combination of external signals--position with respect to egg cell components, gradient of products from other cells--as well as internal signals, such as genes acting in cascade fashion. The hope to unscramble this mess is to study isolated systems of cells capable of differentiation, like teratoma and stem cell lines, and to understand how the function of genes of eucaryotic cells is regulated: two big but by no means impossible orders.

Lecture 26

Animal development

How does differentiation start in a developing animal egg? Two factors play a role: entry of the sperm and cellular movements. The site of entry of the sperm, in frogs and other amphibia, impresses on the fertilized egg a polarity: it appears to determine which part of the egg will give rise to which parts of the organism. Later in development, the formation of specific organs is made possible by movements of cells and layers of cells with respect to one another.

Cleavage

The first steps in the development of the fertilized eggs of chordates (animals with a dorsal spine) consist of several cleavages: mitoses and cell divisions without accompanying increase in volume. Yet, the early stages of development in three typical chordates--amphioxus, frog, and chick--differ in some important ways, mainly reflecting the different amounts of yolk in their eggs. The yolk is the food the embryo uses in its early development.

In amphioxus, a small fishlike animal, cleavages are symmetrical and produce a mass of cells--the morula. Then the cells of the morula move to the

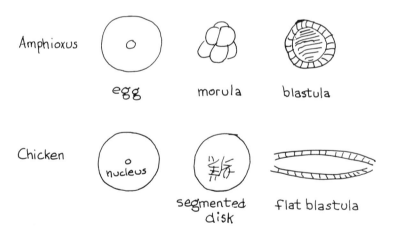

periphery of the egg leaving in the center a cavity full of liquid. This vesicular stage is called a blastula. From this stage on the developing amphioxus embryo must fend for itself. What little yolk there was in the egg has been distributed among its cells, which soon become dependent on food imported from outside.

In the frog egg cleavage consists of unequal cell divisions: the abundant yolk remains in the large bottom cells, which cleave more slowly than the top ones. There is enough yolk to feed the

animal all the way to the tadpole stage. The
tadpole, like the larva of an insect, is an intermediate free-living stage of development between
embryo and adult. It emerges from the egg, grows,
eats, and finally metamorphoses into a frog in
response to a timed hormonal signal.

In the third example, that of the chick, the
egg contains enough yolk to support development
up to the time the chick emerges from the egg.
Cleavage takes place only in a small disk of
cells arising from the fertilized nucleus, which
is located between the yolk and the egg white.
The cell divisions are incomplete and at first
all cells are open to the yolk. In the disk the
embryo develops. Soon some cells begin to grow
deeper and form two layers that are the equivalent
of a flattened morula. A cavity develops between
the two layers, which thus become a flat blastula.

In the chick embryo, cleavage divisions take
place every 60 or 90 minutes; the blastula stage
is reached in one day. (Note that cleavage divisions are faster than those of adult cells in vivo
or in culture; but in early cleavage there is no
growth and little protein synthesis, only subdivi-
sion of existing cell cytoplasm plus mitosis.)

Gastrulation

The next stage of development is gastrulation.
It is the critical one since it determines the
basic design of the adult organism. In amphioxus
and amphibia gastrulation proceeds more-or-less as
if one side of the hollow ball of cells at the
blastula stage had been forced inside, like a soft
rubber ball pushed by a finger. This produces a

two-layered structure. The actual process, however, is not due to pushing but to cellular movements. In a region of the blastula the surface
cells begin to sink toward the inside, giving rise
to something like a circular hole, called the
blastopore. The cells that get to the inside become the entoderm, while the remaining outer layer

is called the <u>ectoderm</u>. The cells around the
edge of the blastopore then start pushing down between ectoderm and entoderm, creating a third
layer, the <u>mesoderm</u>. These three "primitive
layers" exist in all vertebrates and are already
programmed to be the source of specific tissues
and organs: the ectoderm produces epidermis and
nerve cells; the entoderm, the lining of the di-
gestive canal and various glands; the mesoderm,
muscles, connective tissue, cartilage, bones, and
blood. At this early stage of development, therefore, the program is laid out: <u>les jeux sont</u>
<u>faits</u>. Gastrulation has established the pattern
for the complete adult organism.

A special part of the mesoderm will play a
critical role: the one corresponding to the <u>dor-</u>
<u>sal lip</u> of the blastopore (dorsal because on that
side the nerve system will develop). This mesoderm gives rise to a group of cells that are the
precursors of the <u>notochord</u>, the rudiment of the
spine. It is the position of the notochord that
determines the location where a nervous system
will be made by the adjacent ectoderm.

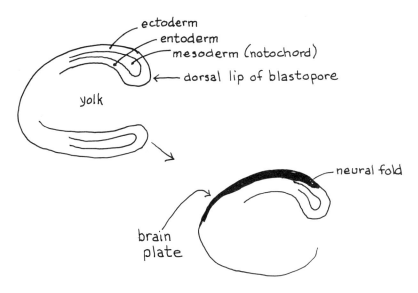

How are these events studied? Essentially in
three ways: by examining sections of developing
embryos with a microscope; by marking regions of
the egg with dyes that can serve to trace the fate
of the dyed cells; and by transplanting portions
of the developing egg. The two latter techniques
can, of course, be used only with free, relatively
large eggs, mainly of amphibia.

What one learns is often quite unexpected.
For example, in the frog eggs the orientation of
the blastopore is not random: it is determined
by the entry of the sperm. At fertilization the
pigment inside the egg moves around, leaving a
gray crescent opposite to the sperm entry site.

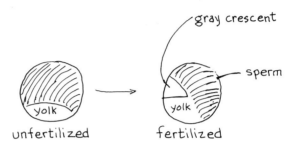

If the first two cleavage cells are artificially
separated, each of them can become a full frog
only if it has at least part of the gray crescent.
(Since cell division has occurred each of the
blastomeres has a nucleus.) The crescent must be
a reservoir of a substance needed for normal
development. Furthermore, the location of the
gray crescent marks the eventual location of the
blastopore and thus is the fundamental locale
that establishes the plan of the mature organism.

Induction

If a portion of the dorsal lip of the blastopore
is removed and transplanted into another embryo
that contains a gray crescent, a double embryo is
produced, provided the transplanted material contained some of the mesodermal cells that had been
programmed to become notochord cells. If the

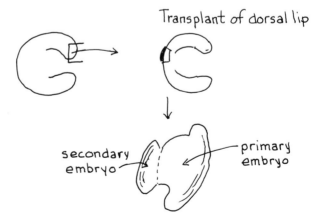

transplanted piece of the dorsal lip is large, a

whole second embryo may be formed; a smaller
piece may produce only an incomplete embryo. But
the presence of some prospective notochord cells
in the transplanted piece is essential because
without them no nervous system develops. If
notochord is present, either in its normal site
or via transplant, the adjacent ectoderm folds up
into a ridge and then a tube, the beginning of
brain and spinal cord. The gastrula has turned
into a neurula. In the transplant experiments,
one can easily verify that there is actual induction of a new structure rather than just a development of the transplant into an embryo. It suffices to transplant the material from a pigmented
species to a light one: the secondary embryo is
a mixture of dark and light cells.

Nerve tissue and notochord together induce the
formation of the somites, that is, of the embryonal
segments from which muscles and bones will arise.
Thus development can be characterized as a cascade
phenomenon, in which the first critical step is
the entry of a sperm into the egg. The dorsal
lip of the blastopore, whose location is determined by the point of sperm entry, has been called
the primary organizer because its presence serves
to organize an embryo. Chemicals of as yet unknown nature are undoubtedly secreted by the
organizer cells and unleash the developmental
process.

Even at this point, however, everything is not
yet settled. The development of each organ reflects the activity of secondary organizers,
which have to be present or activated for the
next step to occur. The nervous system itself,
as already mentioned, acts as an organizer for
other parts of the body. More specific experimental evidence comes, for example, from the
development of the eye in mice.

From the ectodermal nerve tissue of the brain
an eye bud grows out. This then becomes an eye
cup, which later gives rise to the retina. In

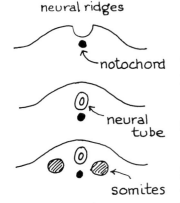

neural ridges

notochord

neural tube

somites

Secondary induction

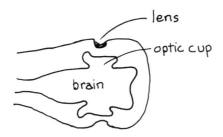

lens

optic cup

brain

front of the eye cup the overlying skin layer begins to thicken and eventually separates from the
rest of the skin and becomes the lens. Thus the
lens, an essential part of the eye, develops in
response to the presence of the eye cup. If an
eye cup is excised and transplanted under the
skin in another part of the body of a mouse embryo,
a lens will appear at the new location. Transplantation of any other part of the brain under
the skin fails to induce lens production. The
eye cup, therefore, is programmed, not only to
develop into the retina, but also to program in
its vicinity the formation of the other parts of
the adult eye. Note that programming does not
necessarily mean intermingling of cells: retina
and lens do not merge. The eye remains a juxtaposition of two different tissues, whose presence
influences their behavior.

In some instances, however, tissues of different
origins actually merge to produce a functional
organ. In the formation of the kidney, for example, morphogenesis must create tubules through
which the urine passes and is chemically modified.
The tubules, as we shall see later, must come into close contact with blood vessels. In the
early stages of the development of the mammalian

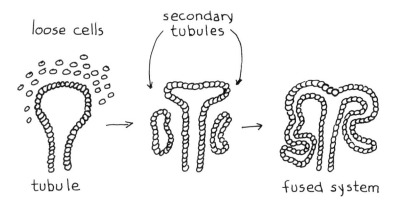

kidney two mesodermal structures meet: a set of
tubules and a mass of loose cells. Upon meeting
the tubules, the loose cells become themselves
organized into tubules, while the tubular mesoderm
branches many times. The two sets of tubules
finally fuse and the complex system becomes organized and functional. Neither the tubules alone
nor the cell mass could develop into a kidney
system. This whole process can be studied in

organ culture: one can place the primitive kidney
bud in a test tube with an appropriate medium.
If either the tubules or the cell mass are removed the development of the other component
stops. What do cells tell each other in these
morphogenetic encounters? We shall look for an
answer in the next lecture.

Lecture 27

Embryonal physiology   The study of the developmental process in vertebrates such as amphioxus, frog, or chick contri-
butes to an understanding of the stages of development of the mammalian embryo. Consider, for example, how the problem of nutrition is solved in
human development. At first the food comes from
the small quantity of yolk contained in the egg.
Then implantation of the fertilized egg occurs.
The embryo derives its nutrition from the womb,
first by simple passage of liquid and later
through a network of blood vessels that develops
in the placenta. Placental circulation permits
exchange of carbon dioxide and oxygen and of low
molecular weight substances, such as inorganic
ions and urea, that can pass from the maternal
to the embryonal circulation. Normally there is
no exchange of large substances like proteins,
but there are some important exceptions such as
the passage of antibodies from the mother to the
fetus.

To become capable of getting oxygen and food
the embryo must create not only its body proper,
but also extra-embryonal organs that suit the
embryonal environment. Just as in flowering
plants, the vertebrate embryo is a different organism than the vertebrate adult.

For example, even before the chick embryo has
completed gastrulation which occurs in a flat
disk, the ectoderm rises to form a ridge, which
ultimately closes all around to form a closed
cavity, the amnion or protective water bag in
which the chick embryo (or the human embryo) remains till birth. At the same time, the ento-
dermal cells grow around the yolk and ultimately
surround it completely. Also, a layer of ectodermal cells, the chorion, grows and will soon
line the tough membrane under the egg shell.

Meanwhile, the entoderm in the embryo proper
has formed the primitive intestine, open to the

yolk. From the intestine there grows out a sac,
the allantois, which will serve the embryo as a
bladder for its urine. (Note: the frog embryo
lives in water and has no such need, and the mammalian embryo can clear its blood of urea in the
placenta.) The allantois grows between yolk and
ectoderm and reaches the surface of egg. Its
mesodermal cells are pushed against the ectodermal

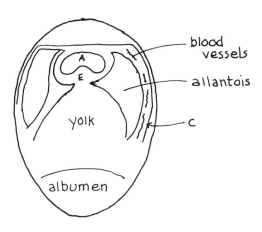

layer. With the allantois there come blood
vessels, which fulfill the function of respiration,
exchanging CO_2 and O_2 with external air through
the shell. These allantoic blood vessels are
functionally equivalent to the placental blood
vessels through which the mammalian embryo performs gas exchanges and gets its food, but of
course the chick embryo gets food from yolk and
egg white. You should note that allantois in the
chick embryo and placenta in the human embryo fulfill the respiratory function of lungs in adult
birds and mammals and of gills in fish.

Morphogenesis

Allantois, amnion, chorion are embryonal organs,
in the same sense as the heart or the kidney are
organs. An organ is an entity with a definite
function and organization and shape, formed in
development through the selected and controlled
multiplication of cells and through the coming
together of different cells by appropriate movements.

The morphogenesis of an organ, that is, the
generation of its ultimate form and structure, involves not only the mutual influences between
different cells during development, but also the
elimination of cells that do not find the proper
location. A good example is that of the germ

cells. In the early chick embryo a small group of cells is set aside in the entoderm that surrounds the yolk as the primitive germ cells, the progenitors of the sperm or egg cells of the mature organism. If this part of the entoderm is excised, no germ cells ever appear. As blood vessels develop, the primitive germ cells continue to multiply, enter the circulation, and are spread to every part of the body. But only those few that reach the site of the genital organs survive and multiply; the others degenerate. Evidently that site is chemically programmed to encourage the establishment of germinal organs.

Thus several mechanisms contribute to morphogenesis: multiplication of cells, their coming together in contact, and active or passive transport of cells to various parts of the embryo. In addition, there is the question of persistence and replacement. Many cells of the body, even in adults, do not last very long, so that new ones have to be manufactured. This permits reshaping of organs. For example, when a bone is broken and reset, the broken parts rejoin forming a callus, which is larger than the original bone; then with exercise the bone gradually returns to its original shape. This is because there are in bone tissue two types of cells: one continuously makes new connective tissue, which becomes impregnated with calcium phosphate to form new bone; the other continuously destroys existing bone. The actual shape of a bone reflects the activities of the two cell types, which in turn are determined by the mechanical forces exerted by the muscles on the bone. Yet the main determinants of the shape of a bone are the genes: in a five-week old embryo bones are already being formed in the shape of adult bones.

In morphogenesis, as in learning, one must distinguish between the effects of nature, that is, the genetic program, and of nurture, that is, the environmental and experiential effects. Usually in morphogenesis the genetic influence predominates, environment playing a relatively minor role. Of course, there are overlaps: an athlete is usually more muscular than a sedentary person; but there are sedentary persons with powerfully developed muscles.

Morphogenetic interactions

What are the signals exchanged between cells in morphogenesis? Consider the skin, for example.

It consists of <u>epidermis</u>, a multiple layered
cellular tissue of ectodermal origin, overlaying
the <u>dermis</u>. This is a mesodermal connective
tissue with a variety of cells including <u>fibro-
blasts</u> which produce <u>collagen</u>, the fibrous protein
that is the structural component of tendons and
bones (and the base of commercial animal glue).

Skin is composed of epidermis and dermis. The
epidermis is continually renewed by multiplication of stem cells in the basal layer and by scal-
ing off of horny dead cells at the surface. If a
fragment of adult or embryonal skin is removed
from a chick and placed in an appropriate nutrient
medium its epidermal cells multiply. If the epidermis and dermis are separated and cultured
separately, however, the epidermal cells fail to
differentiate. If a small piece of dermis or
even a fragment of some other tissue of mesodermal
origin is added to the culture, the epidermal
cells start to differentiate. Thus the epidermis
requires the presence of mesodermal cells. Is
actual contact between the two necessary? You
can find out by inserting some barrier between
the two types of cells in the mixed culture.
If the barrier is porous cellophane, which only
lets through ions and small molecules, the epidermal cells fail to multiply. If, however, the barrier is a filter membrane that allows proteins
(but not cells) to pass, the epidermal cells do
divide. Apparently the message from the dermal
cells is a protein or some other substance of
high molecular weight.

The inducing effect of the dermis is specific
for various parts of the body. Dermis from the
wing induces the formation of wing skin, which
has feathers and no scales. Dermis from a leg
induces the differentiation of epidermis into
leg epidermis with scales and no feathers. Thus
the dermis of one region is programmed to program
epidermis in a given direction. This situation
recalls the one in the early amphibian embryo:
transfer of the cells of the dorsal lip of the
blastopore to another embryo results in the formation of a double embryo.

In none of the above cases do we know the exact
chemical nature of the information that controls
the morphogenetic interactions. Clearly the biochemical information must be specific, but it need
not be a specific chemical. The ratio of concen-

trations of two or more different chemicals at a
given site may be the specific signal. In fact,
even a gradient of one chemical may specify
different signals at different sites if the responding cells have different thresholds for different paths of differentiation. But these concentrations must be graded in orderly fashion, as
in a diffusion gradient. A system of circulating
substances would not do because local concentrations are not specific. Hormones, for example,
reach all parts of the body through the blood.
Their specificity is due only to the specific responsiveness of different target cells and organs.
The developmental signals must have in common
with hormones the target-specific response, but
must in addition be delivered in specified amounts
at the right times to the right places.

**Timing of
differentiation**
  In the course of development cells become programmed to respond to certain signals and other
cells to produce these signals. For example, the
embryo's skin is programmed to respond to the
presence of an eye cup by forming a lens and the
eye cup cells are programmed to induce the skin to
make lens cells. When in development does the
programming take place? Does a cell need to be
already an eye cup cell in order to be programmed
to induce the lens? One way to approach this
sort of question is by using allophenic animals,
a procedure developed by Beatrice Mintz. Allo-

mixed blastula

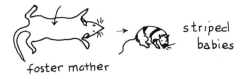

foster mother   striped babies

phenic mice are made by taking embryos from the
womb of female mice in the early developmental
stages, when they are morulae with 8 to 32 cells
and have not yet become implanted in the womb.
The membrane that surrounds the embryo can be dis-

solved with an enzyme so that the individual
cells separate. Then cells separated from two
embryos with different genotypes are mixed together: for example, 32 cells from a black mouse
and 32 cells from a white mouse. If conditions
are appropriate the cells come back together to
form a combined morula. This morula is then
implanted into the womb of another mouse, a foster
mother (which has been appropriately stimulated
with hormones to make her receptive to implantation into the womb). The combined embryo develops
into a mouse. In the case of a combination of
black and white mouse embryo, the allophenic mice
are not gray or dotted: they are striped white
and black. The patterns of stripes are different
on the right and left halves of the mouse. By
counting the stripes in many such mice, one can
conclude that a maximum of 17 stripes can exist
on each side of the animal. There must, therefore, exist just 34 cells, 17 on each side, which
at some point become programmed to give rise to
skin cells. These 34 cells become the progenitors
of the pigment cells of the skin in that segment
of the body.

Let us summarize our picture of development.
At first, in the fertilized egg, there is an
indifferent stage that continues through many cell
divisions. In the course of those divisions, before any differences become visible, there already exists a specific commitment of some cells
as the progenitors of certain other cells, reflecting intrinsic differences among parts of the
egg and sometimes also the asymmetry due to the
entry of the sperm. The fate of a cell may be
due simply to the position where it finds itself
on the surface of the blastula at the time of
gastrulation. Upon gastrulation, a veritable
orgy of chemical interactions between cells begins, so that certain cells in strategic positions
become the masters. As masters, they influence
other cells to become the submasters. A hierarchy
of determination is established that continues to
function throughout development.

Lecture 28

Aggregation

It must by now have become abundantly clear that the central problem of development and morphogenesis--apart from that of differential gene action but directly leading to it--is the nature of the interactions between the parts of a developing organism. We have already seen several instances of such interaction. In this lecture we shall try to explore more closely what is known about the underlying mechanisms. This, unfortunately, is not nearly enough to satisfy even the least demanding molecular biologist.

One way to start is to ask, If one dissociates an organism into its portions, even its individual cells, will these come back together in the proper order? This is reaggregation. Another relevant question is, If one removes an entire section of an organism, can the remaining section replace the missing section? This is regeneration.

Sponges

Let us start with aggregation. The most remarkable findings are those with simple animals such as sponges or coelenterates (jelly fish). Sponges are among the most primitive animals in the presumed tree of animal evolution. A sponge consists of few types of cells--cover cells, flagellated cells that sweep food particles along, contractile cells--but these are precisely oriented and organized for function. (The sponges we use in the bath tub are skeletons of sponges or, more often, poor substitutes made of rubber.) If you take a sponge and separate the individual cells by mechanical or chemical means, then put the cell suspension into a gently rotated flask, the cells come back together into small aggregates and then reconstitute the sponge with the various types of cells in their proper locations.

More interesting still: if one prepares cell suspensions from two species of sponges of different colors and mixes them together in sea water, they reassemble separately, and each population of cells reconstitutes its own sponge. Specific recognition signals must be there to bring about the preferential union of the cells of a given species. These specific signals must be chemical substances on the cell surface or excreted by some of the cells. In fact, some substances present in the fluid of dissociated cells promote reaggregation. These may well be materials of the same nature as those that make various cells of higher animals chemically unique--a subject

Hydra

to which we shall return in a later lecture.

One step up in the scale of complexity is
Hydra: a freshwater relative of jelly fish, a
few millimeters in length, with ectoderm and endo-
derm but no mesoderm (only a "jelly") and a variety
of beautifully organized cell types, including
nerve cells. Hydra is an organism in flux: the
body pattern is maintained, but the constituent
cells change. If you mark a spot of cells below
the "head," they seem to move farther and farther
toward the foot until they are sloughed off.
Likewise, in the tentacles the cells move on to the
tips and fall off. This is because the growth
zone, or hypostome, at the top of the head makes

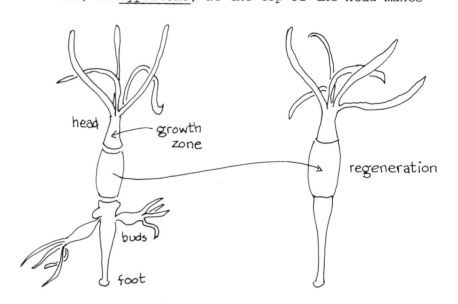

new cells that push away the old ones, generally
without mixing. (The exceptions are the sting
cells, those that sting you when you grab a jelly
fish. They are exploded out and replaced quickly
from new cells moving up from the growth region.)
The new cells acquire specific properties as they
move away from the growth layer. About three-
quarters down from the head is the budding region,
where new heads form, develop, and swim away.

Cutting off the head allows the mid part to
make a new head in the same place. Even a middle
piece without head and foot reforms an animal,
first a head, then a full body and new buds, but
always with the same orientation, head on the
head side. There is strict polarity: each part
of Hydra knows, not only where it is, but also

how it is oriented. A graft of a head onto a cut
middle piece prevents formation of buds until it
has grown away: there is morphogenetic control.

How are these facts explained? In the first
place, there must be influences from the growth
region that direct morphogenesis of other parts
but prevent the formation of buds until their concentrations have been reduced below certain
threshold values. There might be one or two such
inducing and inhibitory substances. If the head
is removed, regeneration may be due to high
levels of induction and/or low levels of inhibition.

An elegant series of experiments by Wolpert
has brought some light on the inhibitory phenomena.
Cut the body of a Hydra (without buds) at various
positions and graft a head on the tail end. Then
cut off the old head (not the grafted one). A

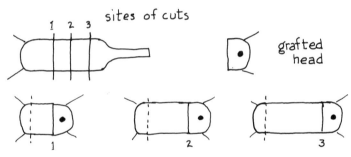

sites of cuts

grafted
head

second cut at various times

Head grafted at 3 must be in place 8 hrs. before second cut
in order to inhibit regeneration of old head. Head grafted
at 1 inhibits even 6 hrs. after the second cut.
←———— diffusion gradient

new head will be regenerated unless the grafted
head from the other end inhibits regeneration,
just as the natural head inhibits buds: what is
observed is that the grafted head inhibits regeneration more successfully the closer it was
grafted and the longer it has been in place before
the second cut was made. This suggests that a
substance from the head must reach the site of
regeneration in adequate amounts. These and
other tests indicate that inhibition propagates
along Hydra at a rate compatible with simple diffusion of a small molecular weight substance made

at the head and lost at the other end. A gradient
of the inhibitory substance is established and
controls the properties of various locations.
(Notice the similarities between these phenomena
and plant growth and morphogenesis, where meristems differentiate according to specific polarities, apical buds control other buds by plant
hormones, etc.)

Gradients and
cellular programming

One question arises: If inhibition alone were
at work, what maintains polarity when the inhibitor
is removed? Why does a thin slice of <u>Hydra</u> know
which way is which? Perhaps cells at a given
level are chemically programmed, not to be individually oriented but to behave morphogenetically in
a specific way. This is in fact true. Gierer has
succeeded in treating <u>Hydra</u> as one treats sponges:
he disaggregated the cells, put them again together in a centrifuge tube, and watched them re-
organize into aggregates, which then form heads,
tentacles, and feet, and ultimately start budding.

Going one step further, Gierer reconstituted
<u>Hydras</u> from their cells <u>in layers</u>, that is, by
centrifuging in various orders cells from head or
from mid-region. Even though polarity had obviously been lost in disaggregating the cells, the new
tentacles always came from the layer of former
head cells, never from the mid-region cells. Not
only that, but even when mid-regions were divided
into two sublayers, the head-side and the foot
side, new heads came from the head-side only.
Notice that these "head-side mid-region" cells

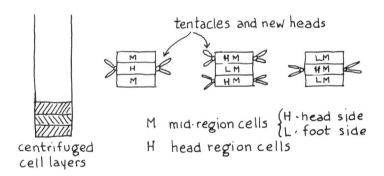

tentacles and new heads

M mid-region cells {H · head side
                     {L · foot side
H head region cells

centrifuged
cell layers

were not head cells, but were more "head" than
the others. Thus cells at a given distance from
the head are marked for distance. They are
stamped, so to speak, with an address corresponding, presumably, to their ability to produce morpho-

genetic and/or inhibitory substance. This marking in turn must have been imprinted by the concentrations of substances to which they were exposed. A good start has been made on identifying at least one of the morphogenetic substances. A small peptide that seems to come from the nerve cells increases specifically rate of head and tentacle regeneration and the number of new buds. This substance is partly free in <u>Hydra</u> cell extracts, partly bound to the cells. Can it by itself explain all gradient properties? We do not know.

   <u>Hydra</u> provides a splendid example of the role of gradients in development and morphogenesis. The gradients here are dynamic: substances are continuously produced and move away by reasons of diffusion only. It has been pointed out by Wolpert that a cell can identify its "place in society" by detecting ratios of concentrations from two points. Wolpert has called this the French-flag problem:

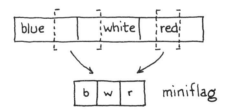

it permits the same pattern to be built on different scales, like a big and a small flag. (Remember the mutant of <u>Dictyostelium</u> that grew into a perfect fruiting body of 12 cells instead of 1,000? The mid-cells must have felt the same ratio of something on either side in the two cases.)

   How unique to sponges and coelenterates are the phenomena of reaggregation we have just described? Embryonic tissues of higher animals offer some examples. Cells of liver and cartilage, dissociated chemically and then reassociated, sort out with liver cells on the outside around a cartilage core. When one uses different animals as sources the cells sort out by tissue (cartilage with cartilage, etc.), not by species. Within the vertebrate world, the markings for "liver" or "cartilage" are stronger than those for "mouse" or "chicken" in producing a decision. The basis of aggregation is probably chemical. An extract of

retina from a chick embryo enhances the reaggregation of retinal cells. But is this substance a
real diffusible signal or is it just a component
detached from cell surfaces?

Specificity signals   In order to promote specific aggregations and
sortings-out in morphogenesis there must be sig-
nals that act at different times and places in
development. You will remember that in the mouse
there is a group of genes, the T genes, that are
needed each at a specific point in embryonal
development, so that any one mutant is arrested
at a specific stage. Substances produced by the
T genes are detected as antigens on the surface of
sperm and other cells and might act in morphogenesis as signals for specific aggregation. Even
more suggestive are certain other mouse mutations
which interfere with morphogenesis in specific
organs: one such mutant, for example, blocks the
relative displacement of certain nerve cell layers
with respect to one another, thereby disarranging
in a fatal way the organization of the brain
cortex. Some normal exchange of signals must be
missing.

As you will see in a later lecture, the morphogenesis of the nervous system has requirements
even more subtle than those for specific aggregation of cells into tissues or of Hydra cells into
a living organism. A given cell from a given
part of the eye retina, for example, must send a
fiber to make contact with a specific cell or
group of cells on the base of the brain. These
cells in turn send fibers to specific parts of
the cortex. To reach its target the incoming
fiber must pass by thousands of cells without
contacting them. The overall plan is genetically
determined, even if developmental accidents may
slightly alter it. What is the secret of the
dogged sleuth work that brings a nerve fiber to
its target?

Steve Roth has done a very pretty experiment.
He took the retinas from early chick embryos, before the nerve cells had started making their
fibers. He disaggregated these cells and labeled
them for a short while with ^{32}P to make them recognizable. Then he mixed them in appropriate ways
with fragments of the tectum, the brain region to
which the nerve fibers from the retina would grow
in the intact chick. It is known that each area
of the retina sends its fibers to a specific

region of the tectum. Roth made mixtures of
cells from various parts of the eye with fragments
from various parts of the tectum in different combinations and discovered that, even at that early
stage, the cells from a given part of the retina
stuck preferentially (about 2:1 or more) to those
portions of the tectum to which their fibers would
later have grown!

If this experiment proves to be of general
validity, it tells us that chemical markings on
the surface of cells can sort out not only cells
of different embryonal tissues, such as liver and
cartilage, but similar nerve cells of the same
organs--the retina and visual brain, for example--
in a way related to their future connections.
Such specificity, which can actually be much
greater at the level of connections between individual cells, is a bewildering reality. What
specific markings can a nerve cell have to distinguish it from a million or a billion other
nerve cells? Ratios of different chemicals, or
geometric patterns of concentration, or of release
of one or many substances? The answer is still
to come.

Part V

PHYSIOLOGY

Lecture 29

The process of development leads from a fertilized
egg to the formation of a mature adult organism.
In plants the adult organism does not reach a
definite size characteristic of the species, but
can continue to grow. Animals, especially mammals,
birds, and insects, have a rather fixed adult size
determined by multigene mechanisms and resulting
in a limited range of variation. Whether the
adult size is gene determined or not, a complex
organism must have elaborate mechanisms to solve
problems, such as excretion, which a single cell
can solve very simply by diffusion of substances
across the cell wall. Thus complex organisms have
organs and organ systems. An organ is composed of
many types of specified cells, which may come from
different developmental pathways and perform different functions.

In order for the cells of different organs to
function specifically, regulation of gene function
is needed, just as in the course of development
regulation is needed to program cells in one direction or another. You remember that in bacteria
the mechanisms for regulating gene function involve
the action of chemical effectors that reversibly
combine with regulator proteins, modulating the
activity of a gene or a group of genes. The same
sorts of mechanisms are presumably involved in
regulation in higher animals, but more complex
events must also be involved in order to explain
the irreversible differentiation of cells whenever
it occurs.

There is one class of chemical regulators whose
nature is relatively well known, even if their
mechanism of action remains by-and-large unknown.
These chemicals are the hormones. Earlier we
saw how the hormone ecdysone stimulates the
formation of the first in a cascade series of
chromosomal puffs during development of the insect
larva. Generally, hormones are substances produced by certain cells of the body and acting to
regulate the function of specific target cells
in other organs or tissues.

Hormones serve many different sorts of regulatory functions in different organisms. In plants,
hormones regulate growth. In insects, hormones
trigger the sequence of events in the development
cycle. In man, hormones control and modulate all
sorts of processes, from sexual development to
calcium deposition in the bones. Many human

Plant hormones

diseases are due to excess or deficiency of
particular hormones. In fact, there may be
several hormones whose existence is not yet known.

Plants, as we have already seen, continue to
grow throughout their life and have no specialized
sex cells set aside early in development. In
flowering plants the main stimuli for production
of sexual organs are environmental: the temperature, the relative duration of day and night. The
nature of the regulatory mechanisms and the specific substances involved are not known.

Plant hormones are primarily growth factors:
auxins, gibberellins, and kinetins, all rather
simple chemicals. The auxins, for example, are
derivatives of the amino acid tryptophan. They
are made by cells in the apex of the plant and
diffuse downward, stimulating the growth of the
apical meristem and inhibiting the growth of side
buds. If the apex is cut off, the stem stops

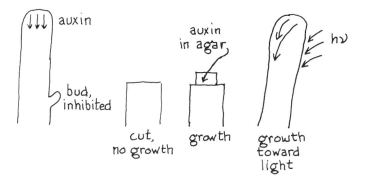

growing, and side buds develop. If the cut-off
tip is placed on top of a piece of agar, auxins
diffuse into the agar. If this block of agar is
then placed on top of the cut tip of a stem, it
stimulates it to grow. If, however, the agar
block is placed below a section of stem, it does
not cause growth. This means that there is a
polarity in the response to auxin--an observation
that recalls the intrinsic polarity of the bodies
of Hydra in their ability to differentiate.

Exposure of a plant to light somehow inhibits
the flow of auxin on the side of the stem nearest
to the light. This is why plants bend toward the
light. [Plants do not respond to gentle talking
or to rock music, contrary to currently widespread
belief. There is no harm done, of course, if
silly people talk to plants instead of pestering

other people.] One can measure the amount of
auxin in a block of agar by putting it off center
on a stem whose tip has previously been removed.
The angle of bending of the stem is a function of

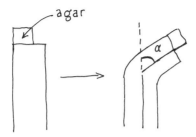

the amount of auxin in the block of agar.

Plant hormones are not hormones in the true
sense of the word, since they do not act specifically on specialized target cells. Rather, they
are growth regulators, which act on growth-compe-
tent cells wherever they meet them. In mammals
there is one hormone, the growth hormone, whose
action resembles that of plant hormones in this
respect.

Insect hormones

In insects, the hormones that control the process of development from the larval or nymphal to
the mature stage are produced by a group of organs
located in the head of the larva. Their role can
be determined by tying off various parts of the
larvae or by making extracts of the various organs
and injecting them into other larvae. The anterior-

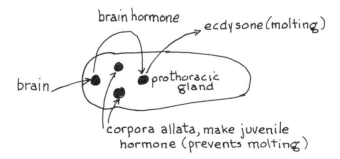

most of these organs produces the brain hormone,
which stimulates the posterior organ, prothoracic
gland, to make and release ecdysone, a steroid
hormone. Ecdysone in turn stimulates the larva to
go through a series of stages of metamorphoses
leading to the production of a mature animal. If
the brain is removed or prevented, by a knot around

the larva, from sending forth its hormone, no
ecdysone is produced, the larva continues to grow
bigger and bigger but fails to metamorphose.

Note that the precise pattern of development
varies from species to species of insects. Some,
such as flies, produce a larva, which then becomes a pupa or cocoon and finally an adult or
imago. Others, such as bedbugs, produce a nympha,
which then goes through a series of moltings and
finally becomes a winged adult. Yet the same
hormones are involved in all insect species that
have been studied so far. This is an important
rule for hormones: they are not species specific
but exert similar roles in reasonably wide ranges
of organisms, for example, all insects or all
mammals. This means, of course, that they act as
triggers to elicit certain activities for which
the cells of given organs are preprogrammed in
development and in evolution. Often the hormone
of a group of organisms, say, primates, is chemically slightly different from the corresponding
hormone of a related group. Do you see why?

Another group of hormone-producing organs in
the insect larva, called corpora allata, make the
juvenile hormone. In the presence of excess
juvenile hormone the larva of a fly or a moth remains a larva, even though it grows in size.
Pupation occurs only when the juvenile hormone is
low. Finally, when no juvenile hormone is present,
the adult tissues develop from the imaginal disks
and the adult emerges. The juvenile hormone,
which controls the destruction of the larval tissues and the development of the adult from the
imaginal disks, is fantastically efficient: one
part in 10,000 by weight prevents maturation of
the larva. The insect never develops completely
and never becomes fertile. Evidently, a convenient
way of controlling various insects may be spraying
with the juvenile hormone. This is actually being
tested experimentally.

Thus in insects the unfolding of various stages
of genetic program is triggered by the activity of
a set of hormone-producing organs. What remains
unknown is the mechanism that controls the activity
of these organs at a particular age of the insect
life. This is comparable to the issue of why
menstruation in women and spermatogenesis in men
begin at certain specific ages, that is, the timing
of the program as distinct from its operation.

Human hormones

Note also the role of the juvenile hormone: the question at hand is that of youth and adulthood and aging.

The many known hormones of man and mammals can be classified according to several criteria: (1) their chemical nature; (2) the organs that produce them; (3) their mode of action. Most impor-

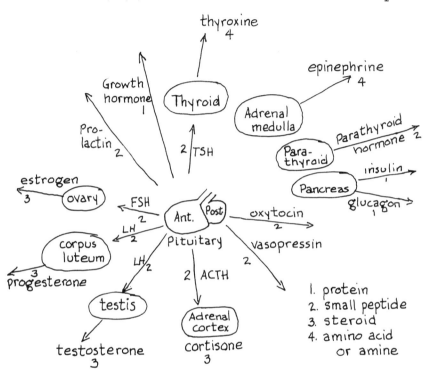

tant is to understand their roles in terms of the expression of the genetic program. Let us start with the thyroid hormone. The thyroid gland is one of several endocrine or hormone-producing organs. It produces the hormone thyroxine. If too little thyroxine is produced, a person becomes metabolically sluggish: he suffers from hypothyroidism. With too much thyroxine, a person has hyperthyroidism: he is nervous and sweats a lot. If the thyroid gland is completely removed, the individual dies unless he continues to take thyroxine.

Thyroxine is a derivative of the amino acid tyrosine, which becomes iodinated by an enzyme in the thyroid. A thyroxine deficiency can occur if there is a malfunction in the thyroid or if there is not enough iodine in the diet. Iodine deficiency in water supplies causes endemic <u>goiter</u>, a

Thyroxine

swelling of the thyroid gland present in many
individuals: the gland tries to compensate for
the lack of thyroxine by making more gland tissue, which is often malfunctional and a source of
various illnesses.

Pituitary hormones   An individual with a normal thyroid receiving
thyroxine for a long period of time gradually
stops producing thyroxine and the thyroid gland
eventually degenerates. How does this feedback
work? The anterior <u>pituitary</u> gland at the base of
the brain produces several different hormones,
which either act directly or stimulate other tissues to produce hormones. All the hormones produced by the anterior pituitary are either proteins
or polypeptides. Among them is a substance called
thyroid-stimulating-hormone or TSH, which stimulates production of thyroxine by the thyroid
gland. Clearly, regulation might occur if thyroxine acted directly on the pituitary gland to
control the production of TSH, reducing it when
the thyroxine level in the body increases. This,
however, is not the level at which the feedback
occurs: thyroxine does not directly inhibit the
pituitary. The anterior pituitary is itself
under the control of a neighboring part of the
brain called the <u>hypothalamus</u>. The hypothalamus
represents the tie between the nervous and hormonal
systems in communicating chemical information to
the body.

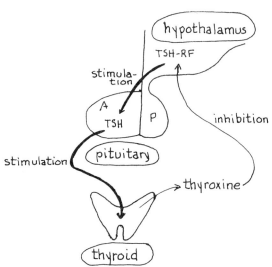

The hypothalamus produces several hormones, including the TSH-releasing factor or TSH-RF. TSH-

RF travels to the anterior pituitary where it
stimulates the production of TSH. It travels by
a <u>portal vein</u> system. The arteries entering the
hypothalamus form a network of capillaries into
which the hormones are released by the cells.
Then the capillaries converge into a system of
veins that leads to the anterior pituitary. There
the veins split again into a bed of capillaries
from which hypothalamic hormones diffuse out to be
picked up by the pituitary cells. At the same
time, anterior pituitary hormones enter the capillaries, which then converge into veins carrying
hormones, including TSH, to the general blood circulation and to various parts of the body. When
TSH reaches the thyroid gland it stimulates the
production of thyroxine. The thyroxine in turn
is carried in the blood both to its various target organs and to the hypothalamus, where it
exerts a negative, inhibitory influence on the
production of TSH-RF.

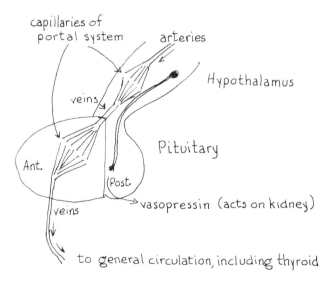

[The portal vein system between the hypothalamus
and anterior pituitary is comparable to the portal
system connecting intestine and liver. Capillaries
and veins from the intestine form the portal vein,
which goes to the liver and there again branches
into a system of capillaries. In this way nutrients absorbed in the intestine are shunted directly
from the intestine through the portal vein to the
liver for chemical processing, before being released into the general circulation. This hastens
the processing procedure: otherwise the unprocessed

nutrients would be diluted in the general circulation and would take more time in reaching the
liver. In general, a portal system performs the
function of specifically shunting blood from one
organ to another. It features a system of capillaries between two veins instead of between an
artery and a vein: the hypothalamus-anterior
pituitary portal system allows almost undiluted
factors to pass from producer cells to target
cells.]

Neurosecretion
Unlike the anterior pituitary, the posterior
pituitary gland is connected to the hypothalamus
by nerve fibers. It is in fact a part of the
nervous system. Some of the hormones (not TSH-RF!)
made in the cells of the hypothalamus are carried
over these fibers and are released in the posterior
pituitary veins. This mechanism is not like other
secretions from cells into blood. It resembles
what happens at certain nerve connections, where,
as we shall discuss later, the excited fibers
release at their terminal endings or synapses some
substances called neurotransmitters. These enter
the adjoining cell and stimulate it or inhibit it.
In the posterior pituitary, however, the nerve
fibers release the transmitter substance into the
blood rather than into other cells, so that the
signal, instead of being passed to one cell only,
can reach many cells.

This kind of generalized neurosecretion is not
unique to the posterior pituitary: it occurs also
in the adrenal medulla. The adrenal gland, located

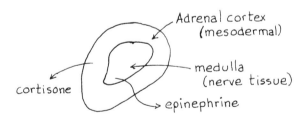

near the kidneys, is a hormone-producing organ
consisting of an inner medulla, composed of nerve
tissues, and an outer layer or cortex composed of
mesodermal tissue. The medulla produces two hormones, epinephrine (also called adrenaline) and
norepinephrine. These are small molecules with
similar structures (they are also called catechol-
amines). Epinephrine stimulates synthesis of
glycogen in various tissues. Norephinephrine is a

catechol   norepinephrine   epinephrine

stimulant of the sympathetic nervous system. In
addition to acting as hormones, these substances
also serve as neurotransmitters at certain nerve
junctions.

The remarkable thing is that in the adrenal
medulla, as well as in the hypothalamus-posterior
pituitary system, a mechanism normally employed
for one-to-one communication between nerve cells
or between nerve and muscle cells has been adapted
to a hormonal action, that is, to an action at a
distance through the blood. Immediately we are
led to consider the reverse possibility: are
there chemical mechanisms of communication similar
to hormones, but released only at the levels of
contact between cells, specific in their action,
and still unknown in their functions and nature?
In the study of development we have encountered
many phenomena that might be explained in terms
of contact induction and of local gradients of
diffusible substances. You will remember that
liver cells, for example, are prevented from
multiplying by something released into the blood
by other liver cells. If liver portions are removed, the remaining cells multiply. The blood
from such a liver-regenerating animal can make
normal liver cells divide. The nature of the control factor is yet unknown. No doubt other
mechanisms of this kind exist in the adult organism
and remain to be discovered.

Steroid hormones   Some hormones are steroids. The steroids have
the general chemical skeleton:

Substituted
positions

The properties of each steroid hormone depend on
small chemical substitutions in a critical area
of the molecule. Presumably these various modifications give steroid molecules affinity for specific proteins of the target cells--but of this we
know nothing as yet. Ecdysone is a steroid; thus
steroids must have assumed regulatory functions
relatively early in evolution. Steroids are
related in structure to cholesterol, a chemical
present in the membranes of all eucaryotic cells,
but never in procaryotic organisms. Steroid-like
substances, therefore, appeared late in evolution
and then assumed both structural functions and
regulatory activities.

There are two main groups of steroid hormones.
One group, secreted by the adrenal cortex, includes a series of substances like cortisone and
hydrocortisone, all with similar functions in the
regulation of cellular metabolism. They affect,
for example, the production of antibodies by
antibody-producing cells and the proliferation
of white blood cells in inflammation. The biochemical mechanism of their action at the level of
cells is unknown, but it is believed that they
act in the cell nucleus. They enter into all types
of cells, but only target cells, that is, those
that respond to the hormone, have a specific protein that combines specifically with a given
steroid molecule and takes it to the cell nucleus.
It is suspected that once in the nucleus these
hormones act as regulatory substances do in bacteria, that is, attach themselves to proteins that
regulate the activity of specific genes. This has
not been proved, but there is evidence that enzyme
levels are altered by steroid hormones in the
specific target cells.

Hormones and the
reproductive cycle

The most interesting aspect of steroid function
is the regulation of the female sexual cycle. This

cycle differs in different types of mammals, but
it has in common a period called estrus, which in
some mammals represents the only period in which
the female is receptive to the male and can be
fertilized. In other species such as man the
female is always physiologically receptive to the
male, but there is a cycle of fertility related to
ovulation and menstruation. The female sexual
cycle in human beings repeats itself every 28 or
29 days throughout a major part of the life span,
beginning at puberty, at age 10 to 12, and continu-
ing until menopause at age 50 to 60.

At the beginning of each cycle a number of egg
follicles in the ovaries begin to swell; usually,
however, only one egg completes development.
(Sometimes two eggs develop and become available
for fertilization, creating the possibility of
fraternal twins.) At ovulation, the developed
egg bursts out of its follicle and descends the
fallopian tube, where, if it encounters a sperm
cell, it will be fertilized. If fertilization
does not occur the egg degenerates. Meanwhile
the uterine mucosa had greatly thickened and
proliferated in preparation for the fertilized
egg to implant. If this does not happen, the
mucosa degenerates and is sloughed off. This is
menstruation. In the ovary, the follicle left
behind by the egg becomes a hormonal organ called
corpus luteum (yellow body), which controls the
proliferation of the uterine mucosa. The corpus
luteum itself stops functioning if there is no
fertilization and implantation.

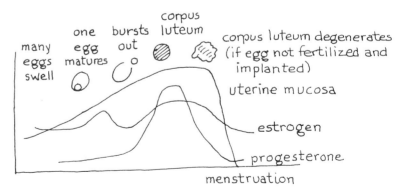

The whole cycle is like a symphony of orchestrat-
ing hormones. First there is a rise in concentration of the steroid hormone estrogen, which is
made mainly by the follicle cells that surround

the egg. After ovulation, another hormone called
progesterone starts rising in concentration as it
is produced by the corpus luteum. Progesterone is
the essential component of "the pill," because its
presence in sufficient amounts in the blood prevents maturation of other egg cells, production
of estrogen, and ovulation.

If an egg is fertilized, it begins cleavage and
when it reaches the uterus it implants itself in
the uterine wall. Then the outer cell layers of
the developing embryo, called chorion, proliferate
and begin to produce large amounts of a progester-
one-like hormone, chorionic gonadotropin. It is
the presence of this hormone that prevents the
breakdown of the uterine mucosa and menstruation.
Therefore in pregnancy the uterine mucosa continues to grow and make itself available for the growth
of the embryo.

Thus the steroids estrogen and progesterone
plus chorionic gonadotropin regulate the whole
cycle. But these steroids themselves must be
regulated in order to achieve a precise cyclic
process. How is their production triggered? Again
the pituitary gland plays the central role. Two
hormones secreted by the pituitary, the follicle

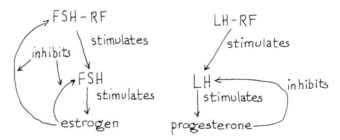

stimulating hormone (FSH) and the luteinizing hormone (LH) control the process through a feedback
loop similar to the one that regulated the thyroid
function. FSH stimulates the development of
follicles in the ovary. When the levels of estrogen and progesterone drop at menstruation, the
hypothalamus is stimulated to produce a hormone,
FSH-releasing factor or FSH-RF. This stimulates
production of FSH by the pituitary. FSH then
stimulates the production of estrogen and follicle
development. When the follicle breaks down and
estrogen production decreases, the hypothalamus is
prodded to produce an LH-releasing factor (LH-RF).
LH-RF stimulates production of LH by the pituitary.

LH stimulates the development of the corpus luteum, which then produces progesterone.

Note that this whole process must be automatic and cyclic, not directional as with insect development. In addition, the process has to lend itself to alteration when the egg is fertilized and pregnancy occurs: chorionic gonadotropin secreted by the chorion of the fertilized egg breaks the cycle. Thus, even though the mode of action of the hormones at the cell level is not known and the nature of the cellular structures that respond to the hormones remains a mystery, the precise organization of the system in space and time is understood. No more can we attribute the 28 day menstrual cycle to the cycle of the moon, as was long believed to be the case.

Testosterone

Another remarkable hormone is testosterone, secreted by special cells (not of the germ-cell line) in the testicles. Testosterone, with other hormones, controls secondary sexual characters: hair distribution, voice depth, general body shape, fat distribution. Castration of a male before puberty causes feminization, and injection of testosterone to young women masculinizes them. Testosterone secretion is itself regulated by the anterior pituitary. As always with hormones, things are not quite so simple as they appear to be at first sight. Women have some testosterones, called androgens, which are made in the adrenals of both males and females. It looks as though the sexual characteristics are the result of a fine balance between varying amounts of estrogens, testosterones, and many other steroids. And, like all fine balances, this one can be tipped one way or the other.

Further reading

Vander, Sherman, and Luciano's Chapters 8, "Hormonal Control Mechanisms," and 15, "Reproduction," offer further information on human hormones and their function.

Lecture 30

Epinephrine

It is clear from what we have discussed up to now that hormonal signals are not directional (except for those that are transferred from hypothalamus to pituitary by portal systems). Most hormones are simply released into the circulation and carried around to all cells. Hence it is the characteristics of the cells that determine the response: during development certain cells are programmed to respond to one or another hormone. In fact, some hormones may produce the same primary effect on most cells where they enter. The specificity of response reflects then not the features of the hormone itself but the cellular reaction to the hormone or to some products of its action.

Take for example epinephrine, the product of the adrenal medulla. When epinephrine enters a cell it increases the production of a simple chemical, cyclic AMP, by stimulating the enzyme that produces cyclic AMP (or cAMP) from ATP (cAMP is the same substance that causes aggregation of Dictyostelium cells; see page 222):

$$ATP \rightarrow cAMP + iPP \; .$$

This action is similar in all kinds of cells, but different cells are programmed to respond differently to increased levels of cAMP. For example, when epinephrine enters liver or muscle cells the cAMP level rises. The level of glycogen, as we shall soon discuss, is regulated by cAMP. When there is need for sugar in the blood, epinephrine and other hormones rise, the intracellular cAMP increases, and glycogen is broken down. Liver cells release the glucose to the blood. When the demand for glucose diminishes, the reverse process occurs and glycogen is accumulated.

In muscles, similar events occur but sugar is not released. It is used in glycolysis to make lactic acid and ATP. The demand for ATP is determined by several stimuli, including electrical stimuli from the nerves and chemical stimuli, that is, epinephrine, from the blood. Epinephrine is in fact a neurotransmitter. The effect of stimulation by epinephrine is the same whether the neurotransmitter comes from the cells of the adrenal gland by way of the blood or is released by a stimulated nerve. The muscle cell has been programmed during development to respond in a specific way to the nonspecific stimulus epinephrine. Other body cells may not respond to the

Muscle contraction

same stimulus at all, or they may respond in different ways.

The discussion has now brought us to talking of muscle and muscle function. Let us first examine the concept of movement. Whether in a bacterium or a muscle cell, motion requires (1) some device that changes its shape or position and (2) a source of energy. In bacteria, for example, there are flagella, protein filaments that can be bent in coordinate fashion to propel the bacterium. Motion of flagella requires energy. In higher organisms, too, motion is associated with protein filaments. In human muscles there are several proteins, at least two of which, actin and myosin, are directly involved in motion.

If a muscle is removed from an animal, hooked up to a suitable instrument, and stimulated appropriately, say, with an electrical impulse, one of two things occur. The muscle may twitch, that is, give a single contraction. Repeated stimulation at intervals generates a series of twitches. If, however, the stimuli are too close together, a continuous contraction or tetanus occurs, because the muscle does not have time to relax and stays continuously contracted.

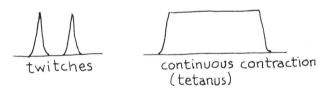

twitches   continuous contraction
                  (tetanus)

Contraction of a muscle can take place in two ways: in isotonic contraction the muscle actually shortens as when you lift a weight with your arm; in isometric contraction the muscle exerts a force but does not shorten, as when you try hard to lift a heavy weight but cannot move it. To measure an isometric contraction it is necessary to measure in some way the force produced. To accomplish these simple mechanical feats, to which no mechanical engineer would give a second thought, natural selection has evolved one of the most exquisite pieces of mechanical jewelry, more refined than the best work of a Swiss watchmaker.

There are two types of muscles, smooth and striated. The smooth muscles in the walls of blood vessels, intestine, and other tubular organs

are contractile fibers that maintain the wall ten-
sion. The striated or voluntary muscles are those
we use for motion or work with the movable parts
of the body. They are called striated because
under the microscope, even without any staining,
their fibers show a striation that turns out to
be the key to their function. The muscle fibers
are not cells--they are fibers made by the fusion
of many cells during development. (This can be
shown nicely by making allophenic mice--remember?--
from embryos that differ in some recognizable
enzyme. The muscle fibers have a mixture of the
two enzymes, whereas single cells have one or the
other enzyme, not both. Do you see why?)

Structure of muscle fibers

A muscle fiber is made up of <u>myofibrils</u>. Each
of these has a banded or striated structure that
looks like this

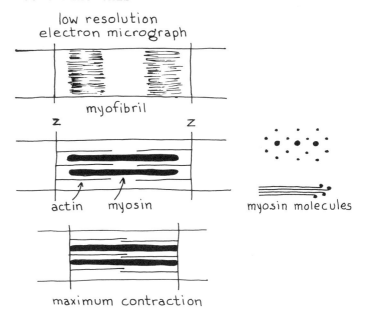

and represents the contractile machinery. It is
made up of two main proteins, myosin and actin,
with several others playing accessory roles.
Myosin and actin are that dream of engineers, an
amazingly efficient chemical engine. Their work-
ings were clarified a few years ago by the
British biologist, Hugh Huxley, in the "sliding
filament" model. The two proteins are arranged
in bundles of filaments so that each filament of
myosin is surrounded by six filaments of actin and
each filament of actin is surrounded by 3 actins

and 3 myosins (remember that in a close packing of
rods each rod is surrounded by six others). The
actin filaments have one end attached to a cross
membrane, the Z-band. Contraction in response to
a nervous impulse, for example, is the sliding of
actin filaments until the myosin hits; then contraction must stop.

The mechanism is even more lovely when one considers the chemistry involved. Actin fibers are
made of thousands of globular protein molecules.
Myosin fibers, on the other hand, are bundles of
filamentous protein molecules, each molecule with
a "fuzzy head." Each fuzzy head projects out
toward one of the six surrounding actins. ATP is
present and binds actin or myosin when contraction
is to take place. Motion is like a ratchet mechanism: a myosin head moves one notch up and an
ATP molecule is split to ADP and phosphate. The
action is cooperative and becomes more so as it
proceeds because more myosin heads become engaged
(those that were in the middle part, without
actin contacts). How the ATP splitting contributes to the sliding is still unknown.

When a muscle acts isometrically, that is,
doing resistant work without shortening, the
mechanism is still the same: ATP is used in
resisting the pull that would make actin and myosin slide away from each other.

You may wonder what happens when contraction
stops. The muscle must return to the resting condition. How is this done? The key to both contraction and relaxation is the movement of calcium ions. An isolated muscle fiber can show contraction and ATPase activity, but only if Ca^{++} is
present. In the intact muscle fiber there are
hundreds of myofibrils side-by-side, all more or
less in phase, and there is a system of membrane
tubules and vesicles that penetrate between
fibrils. This system, called sarcoplasmic reticulum, stores Ca^{++}. Its walls have a pump that
continuously pumps Ca^{++} out of the fibrils and into the vesicles (like all pumps, this one requires
energy--a second use of ATP in muscle). When the
fiber is stimulated, the permeability of the membranes changes, calcium enters the fibrils and
activates the contraction mechanism. When the
stimulus ends, permeability returns to normal,
calcium is pumped out of the fibrils and into the
vesicles, and the muscle relaxes.

This is our first encounter with one of the more remarkable and mysterious phenomena of biology: a selective change of permeability of a membrane to ions in response to a stimulus--a change of electric potential which is itself due to changes of ion concentrations.

ATP and contraction

One more point: Where does the ATP come from? A muscle contains glycogen, which can be converted by an enzyme to glucose-6-phosphate, from where our old friend glycolysis can start and generate ATP. This mechanism, however, is not quite fast enough to provide ATP when the muscle needs it in a hurry. Therefore, the muscle stores about 60 percent of its phosphate, in another substance called creatine phosphate, which like ATP has a high phosphate donor potential. This reservoir of high-energy bonds can regenerate ATP quickly when needed for contraction:

ADP + creatine phosphate $\rightleftharpoons$ ATP + creatine.

Muscle is the only tissue in the body with this special mechanism for prompt delivery.

This supply, of course, is limited: new ATP must be made. The stored glycogen is metabolized to generate ATP and is converted to either lactic acid or carbon dioxide. Lactic acid, as you already know, is produced by glycolysis under anaerobic conditions. This happens in white muscle fibers, which contract fast and have few mitochondria. The lactic acid escapes from the muscle fibers into the blood and is disposed of in the liver. In red muscle fibers, instead, there is plenty of mitochondria; glycolysis occurs under aerobic conditions, the Krebs cycle and electron transport work actively, and the final product is CO_2. You already know that under these conditions there is much more ATP made than under anaerobiosis.

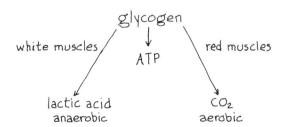

The red muscles contain myoglobin, the red sub-

stance of roast beef juice. You will remember
that myoglobin, a relative of hemoglobin, was the
first protein whose three-dimensional structure
was known in detail. Myoglobin has one heme group
and can pick up only one molecule of oxygen at a
time. Myoglobin, however, has a higher affinity
for oxygen than hemoglobin and takes it away from
it; but it also releases oxygen more readily than
hemoglobin does. Thus the myoglobin in red muscle draws oxygen from the blood, remains oxygen-
ated, and provides a ready reservoir of oxygen for
the muscle.

Glycogen regulation   One substance whose behavior we still must consider is glycogen. Glycogen synthesis and degrada-
tion is controlled by a system of enzymes controlled proximally by cAMP and more distantly by
epinephrine and other hormones. Glycogen is
broken to glucose-1- phosphate

$$glucose\text{-}1\text{-}P \underset{phosphorylase}{\overset{\substack{glycogen \\ synthetase}}{\rightleftharpoons}} glycogen$$

by the enzyme phosphorylase, and is synthesized by
glycogen synthetase. Phosphorylase and phosphatase
are enzymes that can exist in an active and an inactive form. Inactive phosphorylase is converted
to the active form through the action of another
enzyme, a phosphorylase kinase, which attaches a
phosphate (from ATP) to a particular site on the
enzyme molecule. When the active enzyme is to be
inactivated, another enzyme, a phosphatase, removes that phosphate group. This is an example of
a new regulatory mechanism: enzyme activity
regulated by the covalent addition or removal of
a chemical group through the action of another
enzyme. Note that this is different and energetically more costly than the regulation of enzyme
activity by the reversible combination with effector molecules. Here the molecule of the enzyme
becomes permanently changed until the added phosphate group is cleaved. Hence there is no modulation of activity: only a + or - state.

So much for breaking down glycogen when sugar
is needed. The enzyme glocogen synthetase is
responsible for glycogen synthesis to restore the
reserves. A similar control mechanism exists,
only in reverse: a phosphatase converts the syn-

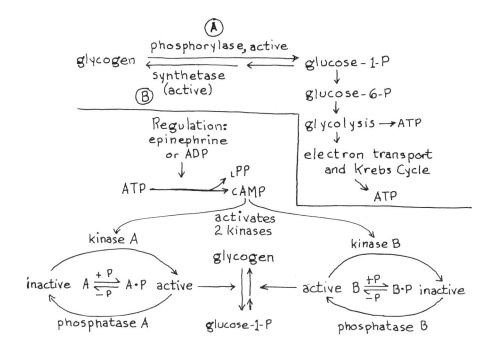

thetase enzyme from its inactive to its active form; and a kinase (plus ATP) converts it from its active to its inactive form.

These two enzymes, one that produces glycogen from sugar and the other that releases sugar from glycogen, are regulated depending on the needs of the cells. When sugar is needed, cAMP stimulates the action of both kinases, causing on the one hand activation of the phosphorylase enzyme and release of sugar; and on the other hand inactivation of glycogen synthetase and inhibition of glycogen synthesis. A single hormone, epinephrine, can produce both effects by increasing the level of cAMP in the cells. When the needs for sugar are low, cAMP levels decrease, glycogen synthesis is restored, and its breakdown is inhibited.

In summary, a muscle fiber is like an engine or, rather, like a machine with an engine fueled by ATP. It has a reserve store of fuel--glycogen; a prompt store of fuel--creatine phosphate; and a store of oxygen--myoglobin. It is fed sugar and O_2 by blood and is exhausted of CO_2 and lactic acid by blood. It needs just a nerve impulse to fire; but that will have to wait a few more lectures.

Cellular fibers and cytoskeleton

At this point, since we have talked of the structure of muscle fibers a digression is in order. Now and then in these lectures we have

encountered fibrous cellular structures of various sorts: the myofibrils of muscles, with their
fibers of myosin and of actin; the fibers of the
mitotic spindles; and so on. What is peculiar
to all these fibrous structures?

The electron microscope has shown that the
intracellular fibers differ in their molecular
organization. Myosin, for example, is a bundle
of linear polypeptide molecules with a more com-
pact end. Actin, however, is a polymer of a
globular protein. The mitotic spindle fibers are

 actin filament

 } microtubule

not filaments but microtubules, that is, tubular
assemblies of a different globular protein called
tubulin. Both actin filaments and microtubules
are supposedly capable of reversible assembly
from monomers. For example, when a cell enters
mitosis the spindle fibers appear, presumably by
assembly, and they disappear as chromosomes
migrate to the poles, presumably by disassembly.
Colchicine, a specific inhibitor of chromosomal
migration, causes disassembly of microtubules.

Apart from mitosis and muscle fibers, microtubules and actin-like filaments (and also thinner
filaments) are present in almost every cell that
has been examined. Although their function is
almost unknown, they appear to form a cellular
skeleton or cytoskeleton--a peculiarly plastic
skeleton that can appear and then dissolve into
its unit components. The most plausible data indicates that these structures are related to changes
in cellular shape and motion, as in amoebas or in
migrating cells. It would be convenient to have
a skeleton that could be folded like an umbrella
when one did not need it.

The important thing is that this cellular
skeleton becomes permanent whenever differentia-
tions of cell shape become permanent, as in muscle
fibers. The most remarkable examples are flagella
and cilia. Within the tail or flagellum of a
human sperm cell, or in the flagella of a protozoan, or in the cilia of a sensory nerve cell
there are regular patterns of microtubules, usually

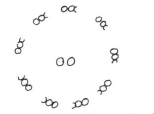

section through centriole
or cilium

in pairs or triplets nine in a circle. In the basal part of a cilium or flagellum there are two additional central tubules. These generally continue within the cells into structures called centrioles present also in the centrosomes at the poles of the mitotic figure as well as at the base of flagella.

There is evidence that the microtubules in cilia are a contractile system, not unlike that of muscles. Centrioles may be primers or organizers for formation of microtubular structures. They may provide an orienting pattern for tubulin to assemble. There is no reason to believe that the microtubules, any more than the filaments, have any liquid-conducting role.

Assembly

One word on protein assembly. You have learned a lot about assembly of protein subunits into molecules, as in hemoglobin. You now have heard about actin filaments, which consist of only one monomer type. But the principle of protein assembly in morphogenesis goes much farther. The shells of certain bacteriophages are made of dozens of different types of protein molecules, each in a specific proportion, thousands of them in one phage particle. All these proteins come together without expenditure of energy and without a guiding template, purely by the contacts between their faces, like molecules building a crystal. In higher organisms there are not many instances of the building of complex structures by assembly of diverse proteins alone (as distinguished from phospholipid membranes). Centrioles may be an example, ribosomes another one, except for the possible role of their RNA. In a ribosome, however, each of the about 80 protein molecules present is different: the result is more like a jig-saw puzzle than a crystal. Yet all these facts add up to one more role for proteins within cells: the provision of a flexible, plastic, and even self-effacing skeleton.

Further reading

Vander, Sherman, and Luciano's Chapter 9, "Muscle," has a clear discussion of muscle function. See also Katz, Chapter 11, "Initiation of Muscle Contraction," for a tie of this subject with that of lectures 33 and 34.

Lecture 31

Blood circulation

In muscle contraction, after all the ATP has been used up, its sources--sugar and glycogen--must be regenerated from nutrients brought in from the blood. In this lecture we shall discuss several other properties of blood and circulation, specifically the actual mechanics of circulation, the mechanisms for transport of oxygen and carbon dioxide, and the mechanisms by which salt and water concentrations are maintained and the products of cellular metabolism are excreted.

Blood is pumped throughout the body by the action of the heart. The heart has a left and a right <u>atrium</u> and a left and a right <u>ventricle</u>.

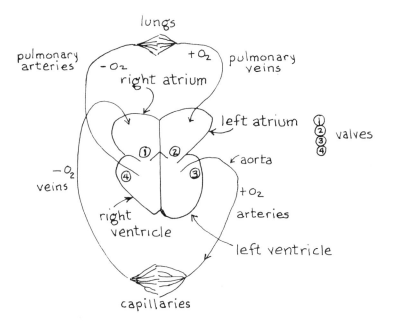

Blood is pumped from the left ventricle into the general circulation. Blood returning from the body passes into the right atrium and on into the right ventricle, whence it is pumped to the lungs. Blood returning from the lungs enters the left atrium and then the left ventricle. As blood passes through the lungs, it exchanges gases with air, which enters and exits the lungs at a rate of about 5 l/min. Blood itself passes through the lungs at 5 l/min. The result is that 200 ml of oxygen passes into the blood and 200 ml of CO_2 is released from the blood each minute. Thus by the time the blood goes back into the general circulation it has a high oxygen concentration, that is, a high partial pressure of oxygen. As blood

passes through the body, oxygen is lost to various
tissues, so that when the blood returns to the
heart the oxygen pressure is 40 mm of mercury.
Since in air the oxygen pressure is 150 mm of
mercury, and in the alveoli of the lung--that is,
in the spaces where gas exchanges take place--
oxygen pressure is 105 mm, the blood becomes re-
charged with oxygen. A similar but reversed situation occurs with CO_2: the concentration is lower
(40 mm Hg) in the blood entering the general circulation than in the blood returning to lungs
(46 mm Hg). In the lungs the excess CO_2 is lost
into the air.

In the blood, however, there is a major difference between the handling of CO_2 and O_2. The
O_2 is practically all carried within the red
blood cells in combination with hemoglobin.
Hemoglobin also has affinity for CO_2, but only
about one-third of the CO_2 is carried by hemoglobin; the remaining two-thirds of the CO_2 is
dissolved in the blood as carbonate. That is,

$$CO_2 + H_2O \rightleftharpoons HCO_3^- + H^+$$

This is a very important reaction catalyzed by an
enzyme in the blood, because the concentration of
hydrogen ions in the blood controls the respiratory
rate. Sensors located in the brain respond to the
blood concentration of H^+: if this increases,
the respiratory rate increases, causing the elimination of more CO_2 from the body. This loss of
CO_2 drives the reaction to the left and brings
about a decrease in the H^+ concentration in the
blood.

Hemoglobin and O_2
transport

There is a remarkable relation involved in the
transport of oxygen with hemoglobin. Hemoglobin
is really a device to make oxygen available to the
tissues as they need it--one might almost say, as
they deserve it. The molecular structure of hemoglobin is especially suitable to this exchange.
A graph of the relation between the partial pressure of oxygen and the percentage of hemoglobin
molecules saturated with oxygen looks like this.
(Note that oxygen binds reversibly to the heme in
both hemoglobin and myoglobin.)

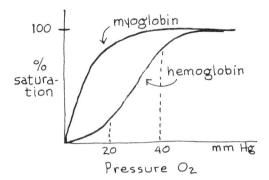

Pressure O_2

The steep part of the S-shaped hemoglobin curve is located between 20 and 40 mm of mercury, that is, just below the oxygen pressure in the blood after it has circulated through the body. In this region of concentrations, therefore, hemoglobin tends to release oxygen in response to small changes of oxygen in the blood.

The shape of the O_2 dissociation curve is related to the structure of hemoglobin. As you know, a hemoglobin molecule is composed of four polypeptide subunits, $\alpha_2\beta_2$. Each of these subunits has a heme group containing an iron atom that can bind oxygen. When a molecule of hemoglobin receives the first oxygen molecule, its conformation changes and as a result the affinity of the other heme groups for oxygen increases, so that additional oxygen atoms are taken up more readily. The molecular basis for this is complicated: essentially the distortion of one of the subunits by O_2 causes a greater avidity for O_2 in the other hemes through stresses propagated among protein subunits. This is a cooperative effect, in the sense that efficiency is increased above what could be achieved by simple summation at four independent subunits.

Consider a muscle, for example. The blood that circulates through it is pushed by the hydrostatic pressure from the heart through arteries and through capillary networks. In the capillaries circulation is very sluggish. Oxygen and carbon dioxide diffuse in and out of the capillaries. The oxygen pressure in the blood decreases and this lowers the affinity of hemoglobin for oxygen. The oxygen in the red cells, therefore, is released into solution and diffuses into the muscle cells until equilibrium is reached, that is, until the concentration of oxygen in the cells

is about the same as that in the blood. Hemoglobin thus acts as a reservoir of oxygen, which
is tapped depending on the pressure of oxygen in
the tissues and therefore in the blood plasma.

Blood composition

Blood itself is composed of plasma, red blood
cells, white blood cells, and platelets. Plasma
(do not confuse with serum) is the clear liquid
that remains after the cells have been centrifuged
away. Serum is the liquid that remains after the
clotting of plasma. In the clotting process,
which we already mentioned, several proteins are
removed, including the enzymes that cause clotting.
Serum does not clot. It contains about 7 percent
protein, less than 0.1 percent glucose, and a
variety of salts. Plasma will clot unless it is
defibrinated, for example, by shaking it with
glass beads. Clotting can also be prevented by
the addition of anticoagulants, substances that
block the action of clotting enzymes. These substances are used in the treatment of patients with
bone fractures or who are otherwise in danger of
blood clotting.

Heart function

What makes the blood circulate? First there is
a rhythmic contraction of the atria that pushes
the blood into the ventricles. Then a contraction
of the ventricles pushes the blood into the circu-
lation--pulmonary circulation from the right ventricle and general circulation from the left ventricle. The rhythmic contraction occurs because
the cells of the heart are <u>autorhythmic</u>: heart
cells, even when separated and placed in culture,
continue to contract. Even embryonic cells that
are destined to become heart cells, if they are
cultured <u>in vitro</u>, begin contracting spontaneously
after a few days. The heart cells are muscle
cells but differ from those of body muscles because they are not fused together and have a single nucleus per cell. To a first approximation,
they would appear to be independent and, in culture, they do in fact behave independently.

In the heart, however, the cells do not contract
independently. In the right atrium there is a
region called the <u>sinoatrial node</u>, which acts as a
"pacemaker" for the whole heart. The sinoatrial

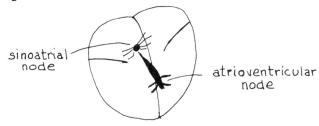

node sends out a signal, which spreads to all the cells of the two atria in about one-tenth of a second; this is much less than the time required for one heartbeat (about one second). The rhythm is kept constant because the heart cells have a <u>refractory period</u> of 0.25 sec, that is, after having been stimulated and having contracted, a heart cell does not contract again for 0.25 sec. This refractory period, like that of nerves, is due to an alteration of ion distribution across cell membranes. The refractory period allows for the contraction to spread and for the cells of the atria to contract almost simultaneously and then wait for the next signal from the pacemaker.

How does the impulse spread to the ventricles? The heart is really two organs: atria and ventricles are separated by a partition composed almost exclusively of noncontractile connective tissue. The important exception is the so-called <u>atrioventricular node</u>, a bundle of specialized heart cells. From this node, branches spread throughout the ventricles. Once a signal from the sinoatrial node reaches the atrioventricular node, about one-tenth of a second after it was generated, it spreads rapidly throughout the ventricles. The contraction of the ventricles thus follows the contraction of the atria by a short interval.

If a heart attack--a transient block of a cardiac artery--damages the sinoatrial pacemaker or its connection with the atrioventricular node, the patient is in trouble. The cells of ventricles, released from controls, may contract anarchically-- the so-called fibrillation. An artificial electronic pacemaker placed in the chest can provide rhythmic stimulation to the heart and help substitute for the damaged pacemaker.

What does it mean to say that heart cells are autorhythmic or that the impulse passes from one heart cell to another? An autorhythmic cell can undergo a change of ionic composition in such a way that the electrical charge on its membrane changes. Generally, all cells are more negatively charged inside than outside. This is due to the ubiquitous presence in eucaryotic cells of a sodium pump, an energy-powered mechanism that continuously pumps out sodium ions, thus maintaining a negative potential (for example about -70mV inside heart cells). In an autorhythmic cell, the

potential in the cell increases spontaneously up
to a certain threshold value, at which point an
action potential occurs, that is, a spike appears in
in the potential difference across the membrane.
The cell "fires."

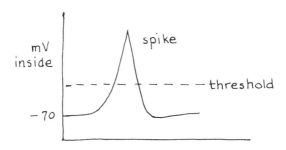

The threshold occurs because at a given ΔV the
permeability of the cell membrane suddenly changes
and allows ions to flow more readily in and out
of the cell. We shall study these phenomena in
detail later in connection with the nervous system.

Tight junctions   In the heart every cell can undergo the initial
slow change rhythmically, and the cells would fire
autonomously, as they do in culture, if they were
not connected. But the stimulus from the sinoatrial node can spread to other cells because the
heart cells, although distinct from one another,
come together to form tight junctions. A tight
junction is a contact so close that ions can go
through and electrical excitations can communicate
directly from one cell to the next. In this way
the cells of the heart are all fixed together in a
single response system because of the refractory
period. After having responded to the signal from
the sinoatrial node, these cells cannot fire
autonomously for a while; then the next signal
comes.

Tight junctions, also called gap junctions, are
so named because of their appearance in electron
micrographs.

The space or gap between the membranes of the two
cells in a tight junction is only about 5 nm.
Ordinary cell contacts are 20 nm or more. In
fact, direct channels probably exist between the
cells in tight junctions, and substances up to
1,000 daltons in molecular weight can go through.

A car engine is arranged so that the motion of
its crankshaft regulates the opening and closing
of the valves. In a similar way the action of
the heart regulates the action of the heart valves.
One heart valve is located between the left atrium
and the left ventricle; a second between the right
atrium and the right ventricle; a third and a
fourth are located at the entrance to each of the
two large arteries leading away from the ventricles.
The valves are opened and closed, not by indirect
controls, but by hydrostatic pressure because of
their structure, which resembles that of the baffle
valves of an air mattress. When the ventricles
contract, the valves between atria and ventricles
are pushed close; meanwhile the valves leading to
the arteries open. When the ventricles relax,
the valves to the arteries close, and the valves
between the heart chambers open.

A major function of blood, besides carrying
foodstuff and hormones from one organ to the
other, is to get rid of materials excreted by
cells. It must do so by transporting them outside the body without changing its own composition.
In fact, it is essential to maintain a precise concentration of many salts and proteins in the blood
plasma under the most varying conditions. There
must be a balance, for example, between the
effects of drinking water (which is hypotonic) and
of eating salt (which is hypertonic). The balanc-
ing occurs not at intake but at the level of excretion, in the kidneys.

Salt-water balance   The problem of salt-water balance must be different for different organisms that live in dif-
ferent environments. The solutions adopted by
evolution are in fact different. There are three
main types of environments: salt water, fresh
water, and air. In animals that live in salt
water, either the blood has the same salt concentration as the water, as in some fishes, or there
is a mechanism for excreting excess salt through
the gills or in the urine. Freshwater fish, on
the other hand, are continuously being flooded
with hypotonic water and must therefore have a

mechanism for excreting the excess water while concentrating salt. Animals that live in air have the opposite problem: dessication. This is solved by covering most body surface with a water-proof skin.

This is another good occasion to point out how wrong the view is that evolution has made every-thing perfect. What could be less effective than the miserable solutions adopted by sea fish, which spend literally most of their food energy pumping salt out of their gills? As usual, the truth is that each species and group of organisms have sim-ply made the best of what they had and managed to survive by the skin of their teeth--or of their gills.

Every complex organism requires a complex sys-tem for getting rid of unwanted materials produced in the course of metabolism and yet maintaining a constant internal environment in terms of water, salts, and osmotic pressure. In vertebrates this function is performed mainly by the kidney, which operates under hormonal control.

Renal function

The task of the kidney is to get rid of things like urea $CO(NH_2)_2$, which is the main discard form of nitrogen in vertebrates and a rather toxic com-pound, without unbalancing the composition of blood plasma or losing useful substances like glu-cose. The solution to the problem is found in the basic structure of the kidney, in the nephron. A

① filtration   ② secretion   ③ resorption

human kidney has about a million nephrons. Each nephron consists of a long tubule, which forms a

cap at one end around a knot of capillaries
(glomerules). The other end leads to a collecting
tube, which carries urine to the ureter and from
there to the bladder. The nephron carries out
three processes: filtration, resorption, and
secretion. The combination of the three solves
the problem we have posed.

Filtration occurs mainly at the glomerule.
Blood flows through the glomerular capillaries.
Plasma minus its protein components passes into
the surrounding tubule cap by passive diffusion
and forms the first urine. Because it lacks blood
proteins, this liquid has a lower osmotic pressure
than the blood (30 mm Hg of osmotic pressure are
due to plasma proteins). As the first urine
passes through the long tubule of the nephron resorption and secretion start taking place. First,
the cells that line the tubule pump sodium ions
out (that is, from inside the tubule to the intercellular spaces and hence to the capillaries that
surround it). This pumping, as usual, requires
energy. Pumping out sodium alone would leave an
excess of negative charge on the inside of the
tube; this is avoided because chloride ions also
pass out. But the exit of these ions increases
the osmotic pressure locally on the outside and,
therefore, water passes out of the tube. In a
human being, the kidneys filter out about 180
liters a day at the glomerules. Of this amount,
99 percent of the water, 99.5 percent of the
sodium, and 100 percent of the glucose are reab-
sorbed to the blood, leaving only about 2 liters
of urine, which, however, contains over 50 percent
of the urea that was filtered out of the glomerules.
Urea is not pumped in and out and only diffuses
according to concentration and osmotic pressure.
(To measure the rate of filtration--180 liters/day,
as mentioned above--one injects into the blood an
easily recognizable, inert substance that is
filtered out in the kidney and fully eliminated.)

For some substances, such as potassium and
hydrogen ions, there is a final mechanism besides
filtration and resorption: tubular secretion.
These ions, whose precise concentration in the
blood is important for many processes, can be
actively secreted into urine that is already in
the tubules.

The existence of all these mechanisms would not
make sense if they were not strictly regulated.

The regulators are hormones, as usual. The hormone vasopressin coming from the posterior pituitary regulates the permeability of tubule cells for water. A steroid hormone, aldosterone, from the adrenal cortex controls the secretion of K^+ and the resorption of Na^+ and Cl^-. As for the sensors in the feedback loops, there exists a set of pressure receptors, the main one located in left atrium of the heart: blood pressure up $\to$ vasopressin down $\to$ H_2O reentry down $\to$ urine volume up. Other sensors specific for various substances are located directly at the level of hormone-producing cells. For K^+, this occurs in the adrenal gland. Sometimes the sensor is at the operating level. The concentration of H^+, for example, is sensed directly in the tubule cells--a minimal feedback loop. Another miniloop regulates iron levels in the body: the cells that line the intestine and absorb iron from food substances sense the iron concentration in the blood. If it goes up, they store iron from the food and this reduces their ability to pick up iron. This can be demonstrated by measuring the uptake of iron by isolated preparations of intestine from animals that had high iron levels. Here it is the intake that is regulated, rather than the excretion. Is this a superior way? Probably not: just one that happened to work well enough.

Further reading

Vander, Sherman, and Luciano's Chapters 10, "Circulation," 11, "Respiration," and 12, "Regulation of Extracellular Water," will give the student more information on blood circulation, renal excretion, and other topics.

Lecture 32

Immunity

The expression of the genetic program during the development of an organism is, in many respects, practically independent of the environment. A human egg produces a human being with all its usual organs before the environmental needs for these organs have been encountered. For example, blood cells and blood vessels develop in the early embryo in response to purely genetic cues and then form the embryonal circulation, which brings CO_2 and O_2 to and from the placenta. Environmental changes can modify the functions of cells and organs, for example, the level of sugar in blood, or even the structure of some organ, as in the hypertrophy or atrophy of muscles due to intense exercise or lack of it. The overall organization, however, is directly under genetic determination. This does not mean that all genetic potentialities of an organism will always be manifested. There are instances in which the genetic program seems to have reserves of variability, destined to cope with a wide range of external circumstances.

The best example is that of <u>immunity</u> phenomena. A human being, for example, has the capacity of producing any one of several thousand different <u>antibodies</u> in protective response to almost any foreign substance that manages to find its way into the blood or into the tissues of the body.

Blood groups

Let us start with the blood groups. If one individual is given a transfusion of blood from another individual at random, trouble may arise. This is because the red blood cells (RBC) have on their surface specific <u>antigenic determinants</u>, called A and B, which are determined genetically. Type A blood has RBC with antigen A specified by gene I^A. Type B blood has antigen B specified by the allelic form I^B of the same gene. The heterozygote $I^A I^B$ has both antigens, A and B. A third allele, i, causes the appearance of neither A nor B (type O blood).

Alleles	Antigens on RBC	Agglutinins in serum
$I^A I^A$ or $I^A i$	A	β
$I^B I^B$ or $I^B i$	B	α
$I^A I^B$	AB	–
ii	O	α, β

If in transfusion a blood containing only A

antigens gets mixed with blood containing only A
red cells (or B with B, or O with any) no problems
arise. However, if blood containing A antigens is
mixed with blood containing only B antigens, there
is a clumping of the RBC's that can lead to clog-
ging of arteries and death. This is because each
type of blood also contains in its serum certain
antibodies, called isoagglutinins, that combine
with those RBC substances that are not present in
the same blood. Thus A blood contains β aggluti-
nins that react with any RBC carrying B antigen;
conversely, type B blood contains α antibodies,
which clump cells with A antigen. Type O blood
contains both α and β antibodies; type AB blood
contains neither. This is clearly an essential
arrangement, otherwise people's blood cells would
clump all the time. But how is it arranged? Let
us first learn more about the system.

Antibodies are proteins; their production is
itself directed by genes. They are made and released by a special class of white blood cells
called B lymphocytes, derived from stem cells
located in the bone marrow. All individuals contain the sets of genes needed to produce α and β
antibodies. Whether α, or β, or both, or neither
is present depends on the RBC antigens that exist
on the RBC's in that individual, that is, on the
genetic structure at the I gene.

The blood group antigens consist of polysaccharides attached to a protein molecule. The three
common blood group antigens, A, B, O, differ only
slightly. The three allelic forms of the I gene
code for variations of the enzyme that inserts
the terminal sugar. The i allele produces an inactive enzyme; the I^A enzyme inserts N-acetylgalac-
tosamine; the I^B allele inserts galactose.

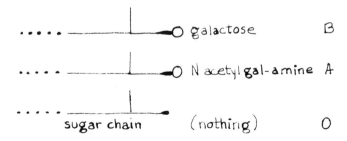

Notice that antibody molecules must exhibit a remarkable specificity to recognize the chemicals just mentioned. And the specificity must be even greater for antibodies that recognize amino acid sequences in proteins than for those that recognize polysaccharide: an antibody may be specific for a given short sequence or group of amino acids and not for any that differ by one amino acid only.

The Rh system

The blood group antibodies α and/or β are normally found in an individual's blood according to his genotype. This is not the case with other antibodies. For example, consider another blood type, the Rh antigen. About 85 percent of all human beings are Rh$^+$: their blood cells contain an antigen that reacts with a particular antibody (which was originally found in the blood of certain Rhesus monkeys; hence the name). Individuals who lack this antigen are Rh$^-$. In contrast, however, to the situation with the α and β antibodies, the blood of an Rh$^-$ individual does not contain the Rh antibody unless he or she has previously been exposed to the Rh$^+$ antigen. The trouble comes when an Rh$^-$ female has a baby fathered by an Rh$^+$ male. If the baby is Rh$^+$, its blood cells may occasionally enter the mother's circulation and induce the mother's body to start production of anti-Rh antibodies. These can then pass through the placenta to the fetus and kill or damage it. Especially serious are repeated pregnancies: with each stimulation the antibodies in the mother's blood reach higher levels and the danger of fetal death becomes higher. (Note: if the mother is Rh$^+$ and the father Rh$^-$, nothing happens!) If an Rh$^+$ male/Rh$^-$ female situation is detected during pregnancy, treatment may consist of drugs that lower the serological reactions of the mother. Usually, such couples should be well advised to have no more than one or two children.

Thus antibodies fall into two classes: those that are produced spontaneously, such as α and β, and those like anti-Rh and many others that are produced in response to the entry of an antigen into an adult organism. Antibodies of the second category are defense mechanisms against infections and their production can be stimulated in order to increase protection. This is the principle of vaccination. But antibodies are produced against any sort of determinants, that is, chemical groupings foreign to the individual.

In the course of evolution the immune system
seems to have been invented at the level of the
higher fishes. It probably evolved in response
to the need to protect animals against infectious
diseases when then moved to fresh water. The
immune system protects the organism against infectious agents because antibodies can destroy
these agents or help body cells to destroy them.
It is a remarkably plastic system since the precise nature of the antibodies that are made depends on the precise antigen that stimulates their
production.

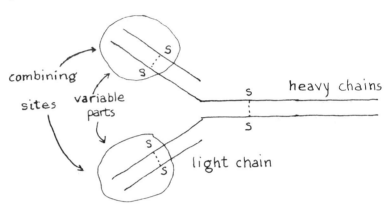

Antibody structure

We already know that antibodies are protein
molecules. Each antibody molecule consists of
four polypeptide chains, two "heavy" chains and
two "light" chains. By sequencing amino acids,
one finds that in any one antibody molecule the
two heavy chains are identical, and the two light
ones are identical. In order to perform the protective function, an organism requires a variety
of antibody molecules sufficient to recognize
hundreds of thousands or millions of different
antigenic determinants. The variety is generated
because one part of each polypeptide chain, heavy
or light, is variable: the rest of the molecule
is constant, that is, identical in all antibodies
of an individual. It is the variable regions that
make up the reactive part of an antibody molecule.
The enormous variety of antibody that each individual can produce reflects the differences in amino
acid sequences in the variable regions. Since the
variable regions of both heavy and light chains
participate in the combining sites of the antibodies, different light chains and different heavy
chains can give many different antibodies. But no

one knows yet whether the variable and constant
regions of a chain are hooked together by gene recombination, messenger recombination, protein-pro-
tein exchange, or some other mechanism.

Given the fact that each individual always produces an enormous variety of antibody molecules,
you may wonder how one could ever get enough of a
single type of antibody to "sequence" the amino
acids of its chains. Any serum is a mixture of
thousands of types. Fortunately, mice (and unfortunately also men) come down with bone marrow
tumors called myelomas. All the cells of a myeloma are descendants of one cell that became
tumoral. Antibodies are produced by one type of
bone marrow cells, and myelomas are tumors of
these cells. Each myeloma produces a single antibody type and provides large quantities of that
particular antibody in pure form for structural
analysis. This is not a special property of
myelomas: any one B lymphocyte and its descendants produce only one antibody type.

How then does the enormous variety of antibodies arise? There might, for example, exist 10^6
different genes, each coding for one specific
antibody, which reacts with one of 10^6 possible
antigens. If so, in each B lymphocyte line, one
out of 10^6 genes could be functional and all the
others not functional (there is enough DNA for
twenty or thirty thousand genes in a human cell).
This mechanism has not yet been ruled out experimentally, but seems rather unlikely. Fewer genes
might be needed if there were, say, 10^3 genes for
heavy chains and 10^3 genes for light chains which,
combined in all possible ways, could give 10^6
antibodies. Another possibility is that an
initially unspecific antibody-forming cell becomes
modified in structure when it encounters a specific
antigen: this would be an instructive mechanism,
the antigen being the instructor. This mechanism
is most unlikely, despite its simplicity, because
the antigen would be expected to remain permanently in order to maintain the specificity of each
antibody-producing cell and all its descendants.
Instead, antigen molecules disappear rather quickly from the body.

Clonal theory
A third alternative is the clonal theory, the
most appealing proposal so far. A clone is the
set of cells descended from one single cell. According to this theory, each of the antibody-producing cells makes only one antibody type because

the precursor cells become programmed in the cour
course of development to make one specific antibody. The programming may be due either to mutations of some special kind occurring in the course
of development of the B cells from stem cells or,
more probably, to a special type of recombination
between a series of several genes corresponding
to the variable regions of the antibody molecules.
A group of n such genes could, by recombination
during development of the stem cells, generate a
great variety of antibody-producing cells. The
mystery of the variable-constant region association
remains.

Genes for variable parts

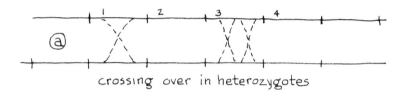

crossing over in heterozygotes

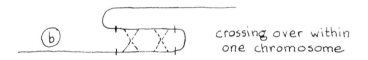

crossing over within
one chromosome

The most important and well substantiated conclusion emerging from the clonal theory is that the
antigen does not alter the nature of the antibody
that a B cell can produce, but stimulates the corresponding cell to proliferate and to generate a
clone, all members of which make the same antibody.
    This brings us to the central problem of im-
munity. What are the antibody-producing cells?
And what role do they or their relatives play in
phenomena other than antibody production?
    The first thing to consider is graft rejection.
If we transplant a piece of skin from a mouse
onto a wound of another mouse, the fate of the
graft depends on the relation between donor and
recipient. If the donor was another mouse of the
same inbred line (almost an identical twin) the
graft takes and heals. If the mice are of a random breed, after a few days the graft is rejected.
The greater the genetic resemblance, the longer
the graft persists. In man, the same rules apply:
self-transplants or grafts between identical twins
ake, the others are rejected. For kidney trans-

plants, to increase the chance of taking, surgeons
prepare the recipient by treatments that lower
immunity responsiveness. There is something
common between antibody response to antigens and
rejection of foreign tissues and organs.

The explanation is as follows. The bone marrow
stem cells can produce precursor lymphocytes. In
the bone marrow the lymphocyte precursors become
B lymphocytes, which can produce circulating antibodies. Other lymphocyte precursors circulate
and reach the thymus, a gland located behind the
upper wall of the chest. There they are converted
into T lymphocytes by contact with some hormone-
like substance. The T lymphocytes have antibody-
like molecules on their surface and are responsive
to antigens, but do not release any antibodies.
Instead, they get stimulated by the appropriate
antigen to proliferate and, if they meet that antigen on the surface of a cell, they attach to it
and destroy the cell. They are the destroyers
that bring about the graft rejection, aided by
other cells called macrophages. If the thymus
is removed from a young animal, the T lymphocytes
do not appear and the immunity mechanism does not
develop normally.

Are B lymphocytes and T lymphocytes two independent developmental forms of the stem cells?
Not quite. B cells do not come into being unless
T cells are present. From experiments in cell
culture it seems that for a B lymphocyte precursor
to proliferate and give rise to a clone of antibody producing cells it must meet both the appropriate antigen and a T lymphocyte. Somehow, the
T lymphocyte helps the B cell to respond to the
antigen. If no T cell is present, the antigen in

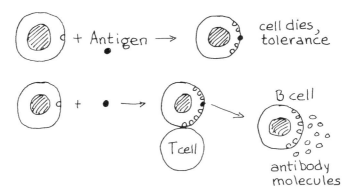

fact may kill the B cell precursor.

This brings us to another set of important, but poorly understood facts. You must have noted in the blood group story that, if a person has red cell antigen A it has no antibody α; if it has blood group substance B it has no β. Convenient-- but how is it done? The answer is tolerance. If an individual--man, mouse, or any mammal--gets massively exposed to an antigen, especially during fetal or early life, it becomes unresponsive to that antigen. The corresponding B cells are killed off or paralyzed. Likewise for graft re- jection: early exposure to a cellular antigen prevents the development of T cells competent for that antigen, and graft becomes more acceptable.

How this tolerance develops is not well under- stood, but it has a very important implication for the biology of individuality. It explains not only why an individual does not normally produce antibodies against its own cells and cell antigens but also why it can accept grafts of its own or its identical twin's organs. Each individual is programmed during development to recognize certain antigenic substances as self and all the others as nonself.

There are certain exceptions that confirm the rule. For example, sperm cells have antigens that can induce B and T lymphocyte response in the same individual. The same is true of the cells that make up the lens of the eye. The reason is that both sperm cells and lens cells, and a few other cells of the body, never enter the circulation, so that their proteins are never met and never in- duce self-tolerance. They are privileged antigens, insulated from the immunity mechanism. There are some diseases in which an individual becomes sen- sitized to produce antibody against some of its own antigens because of extensive release of these from cells and breakdown of tolerance.

In summary, the immune system is one in which a variety of cells, each genetically programmed to produce a given antibody response to an antigen, are stimulated by antigens to proliferate and, if they are B cells, to release antibodies. Both genetic program and antigen are needed to bring about this amplified response. The genetic mecha- nism that creates the variety of antibody genes in the lymphocytes is one of the most mysterious and exciting puzzles of genetics.

One other intriguing puzzle: if specific substances on foreign cells bring about rejection of
organ grafts, why is the embryo, implanted into
the mother's womb and in contact with her blood,
not rejected? After all, about half the embryo's
proteins are like those of the father, hence nonself for the mother. Nature--that is, natural
selection--has provided a special protection: the
cells of the outer layer of the embryo, the so-
called chcrionic trophoblast, are not attacked by
T lymphocytes, probably because they excrete some
protective substance. It has been thought that
cancer cells may also escape destruction by a
similar mechanism.

Further reading   Watson's Chapter 17, "The Problem of Antibody
Synthesis," is a partial view of the subject of
immunology. Most immunology texts are somewhat
obscure due to specialized terminology.

Lecture 33

In a complex organism, in which different functions such as respiration, movement, and excretion are delegated to separate sets of cells, the genetic program must also provide for coordination. In fact, coordination is present everywhere, from the regulation of enzyme synthesis to the spreading of excitation between heart cells. We have already studied one general regulatory system, the endocrine system. The endocrine cells pour stimulatory or inhibitory substances into the general circulation, "hoping" that they will be carried successfully by the blood to the desired target cells. It is targeted only insofar as certain cells are responsive to given hormones.

The other major regulatory system, the nervous system, differs from the hormonal system because it provides specialized communication going from point to point in the organism. It is not a system of communication comparable to a telephone, in which the same channel provides for two-way communication. In the wiring diagram of the nervous system nerve cells are programmed to transmit highly stereotypical signals in one direction only. The receiving cell may either respond or not respond to the signal. If it does respond, its response does not include any encoding of the original signal. My finger bends in a similar way whether it tries to scratch a spot off my coat or to get away from a burning flame. The signals that warn of the fire or of the desire to clean the coat are brought to the muscles by the same nerve. And a given nerve cell, irrespective of the ultimate source of the stimulus, has only the choice between giving or not giving a response. It can vary the frequency of the response in proportion to the intensity of the stimulus, but it does not carry any information as to the nature of the stimulus to which it is responding. A nerve cell does not convey language-type information as a telegraph does when it transmits in Morse code. The nerve message is only firing or not firing. Its meaning is determined by its source and its destination--in other words, by the wiring.

In a mammalian nervous system, there is an enormous number of nerve cells or neurons--10^{11} or more in the human brain and spinal cord. During development some nerve cells are displaced outside the central nervous system (CNS), for example, the smell-sensitive cells of the nose or the neuro-

secretory cells of the adrenal medulla. On the
other hand, the CNS and the nerves have many non-
nerve cells, some of which serve to provide electrical insulation. Neurons are connected in a
variety of ways: some may receive many inputs,
from as many as 10^5 other cells; others may receive fibers from only a few cells. As far as we
now know, all of the billions of nervous connections seem to be genetically determined. The
possibility that practice may create new connections is not excluded, but the most likely explanation of learning phenomena, both in development
and in adult life, is that performance improves
because experience modifies the functioning of
genetically determined connections between nerve
cells, facilitating or hindering the transmission
of signals at the cell-to-cell contact sites
called synapses.

Principles of neurobiology

There are three major questions about the CNS:
(1) How does it work? (2) How does the network
get wired during development? (3) How does it
learn, that is, how does experience change the
function? We have a partial answer to the first
question and know almost nothing about the second
and third. One reason that relatively little is
known about the CNS at the biochemical and molecular level is that, as we shall see, a good part
of the answers lies at the level of the functional
biochemistry of membranes, specifically the inter-
relation between electrical properties and the
permeability of membranes. This is one of the
most obscure areas of biochemistry.

Neurobiology is often regarded as a difficult
or mysterious subject. In part, this is because
our knowledge of it, although exquisitely detailed
in physiological terms, is still incomplete in
terms of biochemistry. Another reason for the
supposed abstruseness of neurobiology is the role
played by electrical phenomena in conduction and
transmission of signals; but this need not scare
the student any more than would the electrical
circuitry of an automobile. To make what follows
easier, it may be helpful, before getting any
further into the subject, to outline some key
facts about nerve conduction, almost a summary of
things to come.

1. All excitations of nerve cells and nerve
fibers are depolarizations, that is, reductions
of existing voltage differences between the two

sides of the cell membranes (usually negative inside when at rest). <u>Hyperpolarizations</u> cause inhibition.

2. Depolarizations and hyperpolarizations can be caused by chemicals released at synapses, or by externally applied electric currents, or by mechanical stretches on membranes.

3. The voltage differences reflect differences in the concentrations of ions, mainly K^+ and Na^+. Generally, K^+ is higher and Na^+ lower inside the cells and fibers than outside. This is due to selective permeabilities of the membrane and to a chemical pump that ferries Na^+ out and K^+ in. Depolarizations and hyperpolarizations result from very small flows of Na^+ and K^+ in and out. When the changes of polarization are caused by chemicals released at synapses, the chemicals act by altering the permeabilities of the membrane to the ions.

4. A fiber fires when depolarization reaches a threshold: then the permeabilities of the membrane to ions change rather abruptly, and Na^+ ions can enter, depolarization reaches a maximum, and the excitation propagates explosively along the fiber (<u>action potential</u>). The original situation is then restored by a return to the original permeabilities and by the action of the pump. The movements of ions in the various phases have been verified experimentally and account quantitatively for the observed changes in potential across the membrane. One can make a good electrical model of the fiber.

5. One remaining mystery of neurobiology is the mechanism by which a membrane is <u>gated</u>: that is, how changes in the electrical potential across the membrane cause selective and nonlinear changes of permeability for various ions. The clarification of these events in terms of protein chemistry is still to come.

This much said, let us now go on to describe the organization of the nervous system. The nervous network is divided into two main subsections: the <u>somatic</u> system, which concerns conscious actions and sensations, and the <u>autonomic</u> system, with a sympathetic and a parasympathetic division, which regulates functions over which there is no conscious voluntary control, such as breathing, heartbeat, gland secretions, and movements of the intestine. For our purposes, however, we shall ignore the differences between these sys-

Neurons

tems since their unit elements are similar.

The cells of the nervous system, called <u>neurons</u>, are single cells, not fused as muscle fibers are. A neuron has a cell body containing the nucleus and a nerve fiber or <u>axon</u> extending out from the cell body. The nervous signals are carried along the axon. A neuron may or may not have <u>dendrites</u>, that is, filaments or lobes that extend from the cell body and are simply functional extensions of it, increasing its receptive surface.

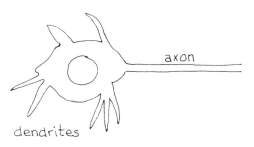

All macromolecular synthesis in a neuron takes place in the cell body; the products of synthesis can then move down the axon all the way to its terminals. An axon may be either <u>afferent</u> or <u>efferent</u>: afferent axons bring signals from the end of the axon toward the cell body and the central nerve organs; efferent axons bring signals from the cell body toward the end of the axon. It is important to realize that this is a "wiring" distinction, not a structural one: given an artificial stimulus, any nerve fiber can conduct in both directions. It is because of the way a neuron is wired, that is, connected at its terminals that its axon does in fact transmit impulses one way or the other but not both.

An axon may at its distal terminal make a synapse with the body of another nerve cell or with many cells if its terminal is branched. It may also synapse with a muscle fiber or a secretory cell. At the synapse stimuli are transmitted from the nerve terminal to the next cell body.

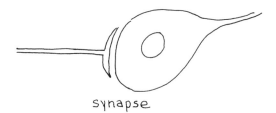

synapse

If an impulse coming from the axon of cell A is
sufficient to activate the next cell B, this in
turn may transmit a signal. But the signal is
changed: it tells only that B has fired, but
does not tell that B has fired in response to a
signal from A. At each station informational content of the signal changes.

Consider, for example, a class of neurons
called <u>Purkinje cells</u> located in the cerebellum.
These are very large neurons that receive at least
three types of connections: from neighboring
cells, from cells deeper down in the cerebellum,
and from cells in other parts of the central nervous system. The firing of a Purkinje cell, the

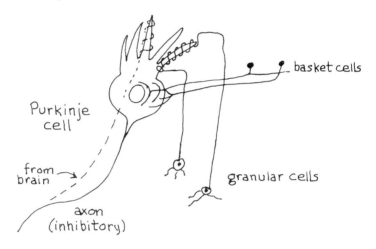

"decision" to send off an impulse, is an integrated response to the sum total of all of the
stimuli that the cell receives from all its connections. Some of these stimuli are inhibitory,
others are excitatory, hence the sum must be
algebraic. The firing of a Purkinje neuron says
nothing about which of the many input synapses
transmitted signals. It only indicates that the
integration of the input signals has resulted in
a net stimulus sufficient to produce an impulse
in the axon of the Purkinje cell. Thus in the
nervous network each cell integrates the signals
it reaches within a reasonably short time interval
and fires or does not fire accordingly. A given
synapse on a given cell can be either excitatory
or inhibitory; but whether its stimulation is
ultimately converted into an inhibitory or excitatory signal depends on what that cell is capable of doing. For example, the Purkinje cells of

the cerebellum have on their body both stimulatory
and inhibitory synapses but they send off only
inhibitory signals because the synapses that their
axons make on other neurons are of the inhibitory
type. The nature of a chemical synapse; inhibitory or stimulatory, depends on the nature of the
chemical secretion at the axon terminals and of
the postsynaptic receptors. More on that later.

Reflex arc

Let us now consider a specific system--the
reflex arc--to illustrate the circuitry of the
nervous system. Take an arm muscle connecting
two bones, one above and one below the elbow. Any
motion at the elbow causes either a stretching or
a relaxation of the muscle fibers. In the muscle
there are special organs called muscle spindles

Stretch receptors

or <u>stretch receptors</u>, which register when the
muscle is stretched. Each stretch receptor consists of a group of specialized muscle fibers surrounded by nerve fibers that come from the <u>spinal
ganglia</u>. These are masses of nerve cells located

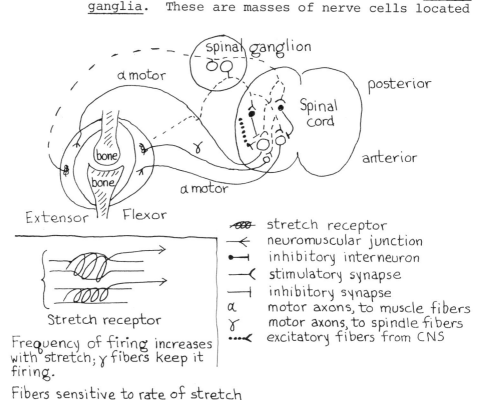

outside the spinal cord, near the holes through
which the spinal nerves emerge from the vertebral
columns. [When I get sciatic pain in my right

leg, it is because the vertebrae press on the roots of the spinal nerves in my lower back; the pain is "referred," that is, interpreted as arising in the leg parts from which the sensory fibers come.] Stimulation of the stretch receptor in the muscle sends an impulse toward the spinal ganglion and to the spinal cord, in which the sensory fibers end. In the gray matter of the spinal cord the sensory fibers make synapses with one or several motor neurons (or α neurons). If the impulse is sufficient, these may then fire and cause a contraction of the muscle and relief of the stretch signal.

This system would evidently be more efficient if the sensory nerve had enough sense to send some terminals to synapse also with some other neurons of the spinal cord that could inhibit the contraction of an opponent muscle--one that extends the elbow and increases stretch in our receptor. This is exactly what happens.

The circuit has one more feature: it is modulated, that is, it is geared to respond quantitatively to quantitative signals. The way this is done is that the motor cells in the cord are integrating neurons, like the Purkinje cells of the cerebellum. They fire when the algebraic sum of stimuli reaches a threshold.

Still one more regulatory feature: as the muscle contracts, the stretch diminishes and the stimulus decreases. If this were too perfect, the response would be minimal. We need a delay system. This is provided by a γ neuron of the cord. This neuron, like the α neuron, is stimulated by the impulses coming from the stretch receptor and keeps firing for a while; its fiber goes to the stretch receptor and keeps its contraction going so that the stretch receptor itself contracts more than the muscle fibers. This positive feedback fools the motor neurons into believing that their action is still needed.

The stretch receptor is our first example of a nervous transducer mechanism: it converts a mechanical signal into an electric signal in the nerve. All sensory mechanisms involve some form of transduction and all transduction, as we shall see, is an interplay of electrical and ionic changes at cell membranes.

The circuit from muscle to spinal cord and back is a reflex arc, controlled automatically by

nerve connections. A voluntary movement of the
same muscle is directed by stimuli coming from
the brain to the same motor neurons in the cord
that are involved in the reflex arc. The motor
outcome is the same because the motor neurons are
the same. If the nerve root entering the spinal
cord from the muscle is cut, the muscle loses the
reflex motion but can still be moved voluntarily.

Electrophysiology

The reflex arc and other circuitry of the CNS
have been explored by the two basic techniques of
neurobiology: neurohistology, the study by light
and electron microscopy of distribution of cells,
fibers, and their connections; and by electrophysiology. Electrophysiology uses microelectrodes,
capillary tubes as thin as 0.4 μm filled with a
KCl solution, to record impulses from within nerve
cells and fibers and also to generate stimuli in
the circuitry. Recording from outside cells or
fibers and especially from outside the skull in
man, although less sensitive can give some ideas
about the features of the system.

A microelectrode stuck into a nerve fiber
measures an electrical potential ΔV with reference
to an outside electrode.

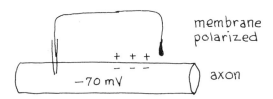

Most nerve fibers have a resting membrane potential E_m of about -70 mV with respect to the outside. That is, there is an excess of negative
charge inside. The nerve membrane is polarized.
When the potential difference decreases, we have
depolarization of the nerve membrane; when it increases, we have hyperpolarization.

Local and action potentials

What happens during nerve activity? There are
two kinds of changes in the membrane potential E_m:
local potentials and action potentials. Local (or
graded) potentials are generated on the surface of
a cell or at a synapse by impulses insufficient
to generate action potentials. The amplitudes of
these potentials are linear with the stimulus
amplitude; they spread slowly, decrease with increasing distance from their point of origin on a
fiber or cell surface, and are summed algebraically

on the membrane. They are the ones, for example, that are received and integrated by a Purkinje cell.

The action potential is different: it is <u>explosive</u>, <u>all-or-nothing</u>, <u>regenerative</u>, <u>self-propagating</u>, <u>self-terminating</u>, and <u>unidirectional</u>.

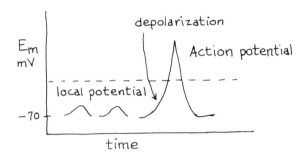

It is <u>the</u> signal in nerves (and also in muscle fibers and heart cells). Its course, as measured inside a giant axon like those of a special nerve of the squid, is like this: The membrane potential E_m rises more or less linearly (it is still a local potential!) until it reaches a threshold; then it rises abruptly to a positive value, about +30 mV, more or less constant for a given fiber. Then it returns to the resting value. The non-linearity, the explosivity of the action potential, is its key feature. At the threshold the membrane permeability to ions changes: the membrane is gated and Na^+ and K^+ gates open and close. We shall see this in detail in the next lecture.

The action potential is regenerative and self-propagating because the local changes in E_m and ion concentrations cause the next region of the nerve fiber to become depolarized; these local potentials again rise above threshold and the action potential moves down the fiber.

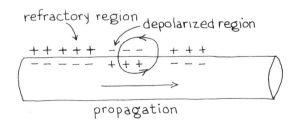

This resembles the explosion in a channel filled

with gun powder: a local input of energy, mechanical or thermal, brings one point to the threshold
temperature and the explosion heats the powder
down the channel to its explosion point. The
initial input of energy releases the stored chemical energy.

The self-terminating feature of the action
potential is essential for it to be useful. The
action potential stops because the changes in
permeability of the membrane are self-limiting;
ions are returned to the original concentrations
and the membrane becomes refractory to stimulation
for a given interval of time. This is the stage
at which energy is used--to restore the resting
potential. You will remember that we already met
a refractory period when we discussed the autorhythmic contraction of heart cells.

The unidirectionality of the action potential
in vivo results from the existence of the refractory period: the impulse can move only where the
fiber is not refractory, and does not last long
enough to stimulate in the back direction.

How fast does the nerve impulse move? The
action potential can go very far very quickly,
as fast as 120 m/sec. The larger the fiber is
the faster the action potential moves (Can you
guess why?). Also, the action potential propagates
faster if the axon is properly insulated, exactly
as for an electric wire. Many nerve fibers are
myelinated, that is, covered by insulating layer
of lipids provided by the membrane of another kind
of cell. The myelin sheath prevents free passage
of ions in and out of the nerve. If an action

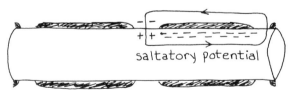

myelinated fiber

potential proceeds down a myelinated nerve fiber,
it jumps from one gap in the myelin sheath to the
next gap; this is called a saltatory or jumping
potential and increases the speed of transmission.

In summary, the nervous system has local signals whose meaning is that they can evolve or add
up into action potentials. Action potentials are

digital, on-off, signals 0 or 1, but also convey
the frequency of the on-off signals. The reflex
arc initiated at the stretch receptor is a good
example: the greater and more prolonged the
stretch is, the greater the frequency of firing
of the afferent fibers but the amplitude of the
action potentials does not change. The maximum
frequency of firing is set by the refractory
period. In the motor neurons, summation of local
potentials occurs and only when the total stimulus
surpasses the threshold value does firing occur,
and the muscles contract as "demanded" by the
stretch.

Further reading   Katz's book is an easily accessible and perti-
nent supplement to this lecture and to lectures
34-36.

Lecture 34

Ionic relations

In the signaling among nerve cells and nerve fibers the membrane potential, that is, the distribution of charges on the two sides of the membrane, changes. Something, therefore, must move: what moves is ions. Their movements, which are electric currents, are controlled by several factors: the existing ion concentrations, the membrane potential, and the permeability of the membrane itself, which is equivalent to a resistance. In addition, the membrane permeability changes as a function of the membrane potential. I have already mentioned that these permeability-potential relations are the key to the nerve conduction and sensory transduction phenomena.

[This lecture requires use of a few notions that students certainly have but may be unaccustomed to use on biological subjects. Let me remind you of the following:

A potential or voltage difference E can send through a conducting system a current A given by E = RA, where R is the resistance (Ohm's law). The current may be carried by electrons, as in a metal wire, or by positive and negative ions. A system that is either nonconducting or not infinitely conducting also has an electrical capacitance C given by Q = CE, where Q is the quantity of electricity stored on a capacitance C at voltage E.

The conductance of a system is the inverse of its resistance. If in a system the current may be carried by several ions, each of these may have its own conductance representing the permeability of the system to that ion. A resistance or conductance can be fixed or variable.]

In order to measure directly the ion concentrations, it is convenient to use the giant squid axon, a nonmyelinated fiber into which one can stick several microelectrodes and do all sorts of tricks. Electrodes inside the fiber measure the characteristics of the <u>axoplasm</u>. Any given ion is subject to electrical forces. At equilibrium the two forces balance and there is no net change in concentration. The equilibrium potential E_{eq} for a given ion, for example, potassium ions, is given by Nernst's equation

$$E_{eq}(K^+) = (RT/F) \ln [K^+{}_{out}]/[K^+{}_{in}]$$

$$= 58 \log_{10} [K^+{}_{out}]/[K^+{}_{in}].$$

Note the analogy with $\Delta G = RT \ln (conc_{out}/conc_{in})$,
the concentration energy (see page 61). F is the
unit of electrical capacity or Faraday. (For negative ions like Cl^- the expression used is $E_{eq}(Cl^-)$
$= 58 \log_{10} [Cl^-_{in}]/[Cl^-_{out}]$. Do you see why?)
    The values $[K^+_{out}]$ and $[K^+_{in}]$ refer to the molar
concentration of the K^+ ion. Typical concentrations actually found under physiological conditions and the corresponding values of E_{eq} are
given in the following table:

	conc$_{in}$	conc$_{out}$	E_{eq}
Na^+	50 mM	460 mM	+55 mV
K^+	400 mM	20 mM	−75 mV
Cl^-	40 mM	540 mM	−64 mV
Organic$^-$	∿400		

The resting membrane potential E_m is −70 mV. Thus,
in the resting nerve, E_K and E_{Cl} are close to E_m,
the resting membrane potential, and only a small
current of these ions tends to go through. [Note:
$E_{eq}(K^+)$, $E_{eq}(Cl^-)$, and $E_{eq}(Na^+)$ are now abbreviated
E_K, E_{Cl}, and E_{Na}.] For Na^+, however, E_{Na} is much
higher and therefore the existing E_m tends to make
Na^+ diffuse into the cell. Something else must
intervene to maintain the resting potential against
the Na^+ drive.

Hodgkin-Huxley theory

    The mechanism of the resting and action poten-
tial in terms of ion fluxes is explained by a
theory put forward by Hodgkin and Huxley. Though
not actually proved in every detail, the theory
has stood up well for many years. It states that
the membranes at rest are almost impermeable to
Na^+, somewhat permeable to K^+, and fairly permeable
to Cl^-. The membrane potential is due almost
exclusively to the internal K^+ concentration. An
action potential occurs because, as the membrane
becomes depolarized its permeability to Na^+ increases (gating!). Some Na^+ enters, the membrane
potential E_m increases and would tend to become
similar to the sodium equilibrium potential
$E_{Na} = +55$ mV. But two other things happen: the
permeability to Na^+ starts going down and the permeability to K^+ increases so that some K^+ can flow
out of the fiber. Therefore E_m decreases back
toward its resting value. (The permeability to Cl^-
does not change.) Finally an energy-requiring

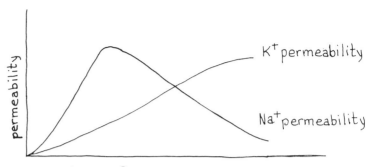

time after depolarization

pump is activated in the membrane and removes the excess Na^+ and pumps in some K^+. For every two K^+ brought in, the pump removes three Na^+:

Outside

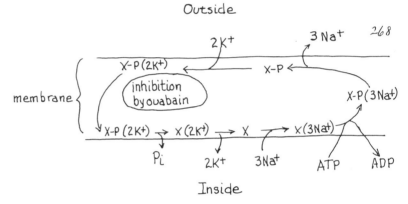

Inside

Electrogenic Sodium-Potassium Pump

Notice that this is an <u>electrogenic</u> pump: it uses energy in the form of ATP to increase an electric gradient (2 charges versus 3) as well as a concentration gradient across the membrane. Thus the membrane potential E_m returns closer to E_K(-75 mV) and, therefore, to the resting potential of -70 mV. (A similar ATP-activated pump has been well characterized in the membrane of red blood cells.)

To understand things a bit better, let us remember that any one ion species, Na^+ or K^+ or Cl^-, is subject to two sets of forces due to different potentials: for K^+, for example, the concentration potential RT log $[K^+_{out}]/[K^+_{in}]$ and the electrical potential, consisting of E_K and E_m. When E_m is at E_K the system is at equilibrium, and no change in K^+ concentration occurs. But if E_m is higher (more positive) than E_K, some K^+ ions tend to move

out due to both the electrical and concentration
gradients. Depolarization, by any cause whatso-
ever, does exactly that: there is less holding
of K^+ ions inside and some of them escape. The
resting potential stems essentially from the unequal distribution of K^+ on the two sides of the
membrane.

As a result of depolarization, the permeability
to K^+ increases and more K^+ ions flow out. But
the permeability to Na^+ increases even faster, and
Na^+ starts coming in under its own gradient. This
is the immediate cause of the action potential,
because the entry of Na^+ more than compensates the
loss of K^+. The reason why the action potential
stops rising is that another change takes place:
a slow effect of the depolarization, which decreases the permeability to Na^+ until it returns
to its initial value. Then the pump has a chance
to remove Na^+ and bring in K^+, and everything returns to the original state.

[It may be useful at this point to bring in a
hypothetical molecular model of membrane permeability changes to help clarify one's thoughts. Even
though it goes beyond what is known today, the
model may be a profitable guide for reflection.
Think of the permeabilities for K^+, Na^+, etc. as
channels provided by protein molecules, singly or
in groups, immersed in the lipid bilayer of the
membrane. Imagine that changes in E_m, that is, in
the ratio of + and - charges on the two surfaces
of the membrane, distort in a reversible way the
protein molecules that act as channels, increasing or decreasing the rate of passage of the ions.
Then it is easy to see, on the basis of what we
know about proteins, how a channel can be squeezed
or widened as the pattern of electrical changes
around the protein molecules changes. The sodium
channels, for example, might first be rapidly
widened by a membrane depolarization and then
slowly constricted by the depolarization if different parts of their protein molecules are distorted at different rates.]

We can now represent the membrane by an electrical model:

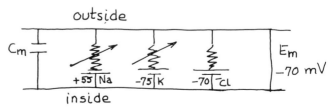

There are three batteries; one, E_{Cl}, is constant
since Cl^- permeability does not change; two are
variable, E_K and E_{Na}. And there is a membrane
potential E_m (and a membrane condenser with capacitance C_m; see below). The changes in permeability affect the current carried by each ion. If
$E_m = E_K$, for example, K^+ does not move. If we
depolarize, that is, make E_m more positive, K^+
ions tend to go out; and so on. When E_m changes,
so do the permeabilities to Na^+ and K^+--or, what
is the same thing, the conductances g_{Na} and g_K,
which are the inverse of the electrical resistances.
The resulting currents I_{Na} and I_K can be represented by

$$E_m \uparrow \rightarrow g_{Na} \uparrow \rightarrow I_{Na} \uparrow \rightarrow E_m \uparrow$$
$$\text{positive feedback}$$

$$E_m \uparrow \rightarrow g_K \uparrow \rightarrow I_K \uparrow \rightarrow E_m \downarrow$$
$$\text{negative feedback}$$

The effects on E_m are opposite, of course, because
the Na^+ current is inward, the K^+ current is outward.

You may be puzzled by the fact that ions come
in and out and change the membrane potential E_m;
and yet, we consider E_K and E_{Na} as practically
constant (as given on page 334). The reason is
that all the big rush of ions in and out involves
only very small numbers of ions. These small
changes in concentration are sufficient to charge
the membrane potential because the membrane is
like a condenser of very low capacitance C. In a
squid axon, for example, there is about 10^{-5} moles
of K^+ within the volume covered by 1 cm^2 of membrane. The amount of ions that need to be displaced to produce 100 mV of depolarization is only
10^{-12} moles, a negligible change in total concentration.

[If you wish, think of the membrane as a set of
bars, oriented across a barrier, and each capable
of taking up a plus or a minus charge at each end.
The bars are large and few, so it takes only a few
charges to change the overall ratio of charges on
the two faces.]

Voltage clamp

What experimental evidence is there in support
of the Hodgkin-Huxley theory? Even with the giant

squid axon it is impossible to do direct measurements since one cannot keep the current or the conductance constant during an action potential. The
key experiments are performed under fixed voltage,
that is, with the <u>voltage clamp</u> device. A squid
giant axon is connected via electrodes to a system
that measures currents and to a special amplifier
that maintains a constant potential.

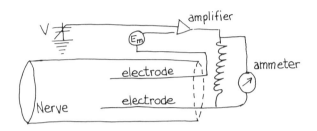

Suppose one sets the voltage at the resting potential -70 mV. Then everything is in the resting
condition. Now you raise the voltage to -15 mV,
that is, above the threshold level: an action
potential would normally start. However the
amplifier immediately feeds in enough current
through a silver wire electrode to maintain the
potential at -15 mV and does not allow the action
potential to develop. The theory predicts that,
at -15 mV membrane potential, the permeability to
Na^+ should be increased; there will, therefore, be
a current of Na^+ ions moving into the axon. The
current needed to balance this sodium current and
to maintain a constant potential should be a measure of the Na^+ flux.

The time pattern of current changes measured
following a shift from -70 mV to -15 mV, for
example, is like this:

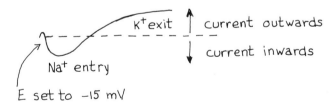

The first blip is the very rapid charging of the
membrane capacity to the new E_m value; the dip is
due to entry of Na^+ as the permeability to Na^+
increases rapidly; the rise is due to K^+ exit.
One can verify all this in a variety of ways. If

external Na$^+$ is decreased, the dip increases. If
external and internal Na$^+$ concentrations are made
equal, the "Na$^+$ dip" is missing. If the inside
of the fiber is filled with radioactive ^{42}K$^+$, one
can measure the tiny amounts of this radioactive
ion that come out of the cell: they correspond
quantitatively to what is interpreted as the K$^+$
current, the slow rise after depolarization.

As far as the nerve fibers are concerned, all
phenomena, therefore, can be interpreted on the
basis of ion currents and permeability changes.
But neurons have two other essential portions:
cell bodies and synapses. The cell body, as you
already know, receives stimuli from many synapses,
and performs algebraic summation of local potentials, which may or may not initiate an action
potential in the axon. Remember that local potentials, below threshold, decrease exponentially
with the distance from the point of origin. At
synapses there are electrical effects and chemical activities. The terminals at chemical synapses may store and release either stimulatory or
inhibitory substances, which can depolarize (excitatory) or hyperpolarize (inhibitory) the postsynaptic membrane. Electric synapses have
specialized gap junctions between fibers and
cells (about 5 nm or less) so that depolarization,
arriving as an action potential, has a high chance
of passing more or less unchanged to the next
cell, in the same way as it happens in the heart
cells. Electrical synapses are always stimulatory
because what is propagated is a depolarization.
This does not mean, of course, that the distant
effects of stimulating an electric synapse cannot
be inhibitory: the receiving cell may form an inhibitory synapse with the next cell. The meaning
of an impulse in a network of nerve cells may
change at every station depending on the nature
of the cell that fires.

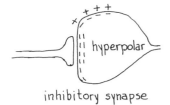

inhibitory synapse

Lecture 35

It may be possible for some smart computer scientist to conceive some alternative system for conducting informational signals in the body that could function at least as well as nerves do. Alternative methods of performing the network functions of the brain, however, are more difficult to conceptualize. The communications between cells at synapses are remarkably efficient because they are both graded and directional. You already know that synapses fall into two classes: electrical and chemical. In an electrical synapse the adjacent cells are partially fused together in gap junctions so that ions can diffuse from one cell to another and propagate depolarization. This is similar to the transmission of contraction signals throughout the heart by means of gap junctions between the heart cells.

Chemical synapses have been studied more thoroughly than electrical synapses. A good example is the neuromuscular junction, in which the ending of a nerve fiber is located in proximity to a muscle fiber, but separated from it by a relatively large gap of 20 nm or more. (Protective cells, called Schwann's cells, similar to those that form the myelin coating of nerves, often surround these synapses.)

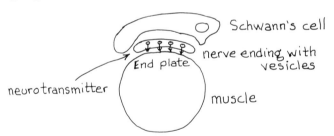

Schwann's cell

nerve ending with vesicles

End plate

neurotransmitter

muscle

Neurotransmitters

The arrival of a nerve impulse stimulates the secretion of a specific substance, a neurotransmitter, into the gap between the ends of the nerve and the muscle fiber. There are different neurotransmitters; the one at the neuromuscular junction is acetylcholine. The neurotransmitter diffuses across the gap, combines with specific receptors on the muscle membrane, and stimulates the muscle fiber, that is, causes depolarization of the muscle cell membrane. Each nerve impulse causes the release of a certain quantity of acetylcholine. In the gap between the nerve ending and the muscle fiber there is an enzyme,

acetylcholinesterase, which splits the acetyl
group off from acetylcholine. Thus we have an
automatic regulatory mechanism, which sees to it

Acetylcholine $CH_3 \cdot CO \cdot CH_2 \cdot CH_2 \cdot N^+ (CH_3)_3$

Norepinephrine HO ⬡ $CH \cdot CH_2 \cdot NH_2$
                        OH OH

Epinephrine   "   " $\cdots NH \cdot CH_3$

γ-Aminobutyric acid $H_2N \cdot CH_2 \cdot CH_2 \cdot CH_2 \cdot COOH$

that the acetylcholine released as a result of
one nerve impulse does not stay around and continue to stimulate the muscle.

The release of acetylcholine from the nerve
ending requires Ca^{++} ions. But this is not simply
a diffusion of the transmitter out of the nerve
terminal: it is a quantal release of a packet of
acetylcholine, presumably in one of the vesicles
that fill the nerve ending. Such packets of
acetylcholine are released spontaneously all the
time, producing miniature potentials of less than
1 mV. Stimulation of the nerve causes the release of several hundred packets in a very short
time.

Acetylcholine acts on the muscle fiber receptor
or endplate by temporarily increasing the permeability of its membrane to Na^+ and K^+. Since the
permeability of the muscle membrane to K^+ is
already fairly high, its change has little effect
on the concentration of K^+ inside the muscle
fiber. But the permeability to Na^+ is normally
quite low: the increased permeability allows
significant amounts of Na^+ to cross the membrane
inward. You can guess the result: a depolarization of the membrane sufficient to produce an
action potential in the muscle fiber and its contraction.

Would acetylcholine stimulate an action potential at parts of the muscle other than where it
synapses with a nerve? Since in some muscles the
nerve endings are located only at the ends of the
fibers, it is easy to try and apply acetylcholine
to other parts of the muscle. The answer is that
acetylcholine does not provoke any action potential except where the muscle fiber synapses with
nerves.

How does the action potential generated in the muscle fiber cause a contraction of that fiber? You may remember that muscle contraction occurs by the sliding filament mechanism and depends on the concentration of Ca^{++} in the muscle fibers. Ca^{++} ions are pumped into the sacs of the sarcoplasmic reticulum within the muscle fibers and stored there. The action potential increases the permeability of the sacs to Ca^{++}, so that Ca^{++} diffuses out of the sacs and into the muscle, where it stimulates contraction, supposedly by activating the ATPase activity of the actin-myosin complexes. After contraction, Ca^{++} is pumped back out of the muscle fiber. One gap in the theory is how the action potential propagated in the outer membrane of the muscle fiber changes the permeability of the inner sacs to Ca^{++}--one more mystery.

The neuromuscular junction is an excitatory synapse: an impulse from the nerve causes depolarization and stimulation of the muscle. Inhibitory synapses are those in which an impulse from the nerve ending makes it more difficult to stimulate the next cell, muscle or neuron: that is, an impulse causes a postsynaptic hyperpolarization. The neurotransmitters in inhibitory synapses are γ-aminobutyric acid and norephinephrine. A given synapse is either excitatory or inhibitory, never both.

Whereas peripheral nerves make synapses with muscle cells or other effectors, in the central nervous system the nerve cells are connected by synapses to form networks. The key point of the networks is that one nerve cell can have both on its cell body and on its dendrites many synapses formed by nerve fibers originating in different parts of the body. Some of the synapses may be excitatory, some inhibitory; some chemical and some electrical. The processing of the inputs from these various sources, as you already know, is an algebraic summation of the local potentials that spread along the cell membrane from the individual synapses. This spreading is not self-sustained; the local potentials decay exponentially

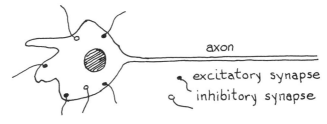

axon

excitatory synapse

inhibitory synapse

with distance from the synapses where they started.
Thus the integration takes into account both the
amount and the location of each stimulus. If the
integration produces a sufficient depolarization
at the site where the axon departs from the cell
body, an action potential is produced. Once this
occurs, however, the only information that is
passed on is that the nerve fiber has fired and
how often it has fired. No information is retained about the pattern of impulses that the cell
had received and which generated the action potential. And yet, despite the stereotyped response
of a nerve impulse, the network, that is, the
pattern of specific connections between neurons,
is all important.

I am sure some of you have already thought of
one interesting feature of the central nervous
system: in computer language, the CNS uses both
digital and analog elements. At every chemical
synapse a digital signal, the frequency of firing,
is converted into an analog signal, the amount of
a transmitter, or vice versa. This dual quality
of the nervous network provides flexibility. Remember, however, that in all cases the basis of
the elements is chemical: it is the gating mechanisms of excitable membranes that convert chemical
disturbances into the all-or-none propagating signals of the action potential.

Nerve regeneration   How is the nerve network created during development? Something can be learned by studying the
process of innervation in the developing embryo.
For example in embryonal muscles, before innervation has occurred, muscle fibers produce an action
potential in response to acetylcholine stimulation
anywhere along their length. After the muscle becomes innervated the muscle fibers become acetyl-
choline-sensitive only at the neuromuscular junctions. If a motor nerve going to a muscle is cut,

the fibers in the end that is no longer connected
to the cell bodies located in the spinal cord degenerate and cease to be excitable. This agrees
with much evidence that the central office for

cellular metabolism of the neuron is its cell body.
As the fibers degenerate, the muscle regains overall sensitivity to acetylcholine. The denervated
muscle fibers twitch continuously in response to
slight variations in their environment. Clearly
it is the nerve fiber synapsed to the muscle that
normally prevents this uncoordinated activity.

If one can prevent scar formation at the end
of the severed nerve the cut nerve fibers grow
back. As they regrow, the nerve endings put out
many fibrils, as though they were searching for
the correct address. Eventually the nerve fiber
grows back to its original location and innervates
the same synapse. (Note how remarkable this is!)
When this occurs, the general sensitivity to
acetylcholine in the muscle disappears, first at
points distant from the synapse and then gradually
closer and closer to it. We can conclude that the

loss of local response to Ach

muscle always makes some substance that produces
acetylcholine sensitivity and that the action of
that substance is inhibited by the nerve. Thus,
besides transmitting impulses, the synapse controls the response of the whole muscle fiber to
the neurotransmitter.

Several chemicals alter the sensitivity of
acetylcholine synapses to acetylcholine. For example, botulinus toxin prevents release of acetylcholine in the synapse; curare blocks all responses
to acetylcholine. Other drugs inhibit the action
of acetylcholinesterase. It seems likely, therefore, that the events at the synapse itself may
be modulated locally by the chemical environment
in the muscles.

How does a regrowing nerve fiber find its way
to its original terminus? Perhaps there are chemical signals waiting there, which are not found in
other locations. But first, how general is the
phenomenon of regrowth and functional recovery
after nerves are damaged? Observations by neuro-
surgeons suggest that the human central nervous
system is quite plastic, that is, relatively capa-

ble of restoring a damaged function by the activity
of another part of the CNS. On the other hand,
specific experiments by the neurobiologist Sperry
on amphibia and mammals have revealed very little
plasticity in the central nervous system. For
example, if one leg nerve is severed and allowed
to regrow, the leg regains its normal movement.
If, however, two leg nerves are severed and then
criss-crossed and allowed to regrow, movement
returns but the front leg, for example, now moves
in response to stimulation of the back leg and
vice versa. The original programming for nerve
connections has been maintained, even though the
fibers have been forced to find new terminals.

If the leg bud is transplanted from one tadpole
to the back of another tadpole, it produces in
the adult frog a fifth limb, which becomes innervated not by leg nerves but by nerve fibers from
the sensory nerves of the back region. When the
extra limb is stimulated, the frog contracts the
hind limb of the same side! Here something new
has happened: the dorsal sensory nerves have
"sensed leg," so to speak, and have made the
appropriate connection in the central nervous
system. Some kind of programming already existing

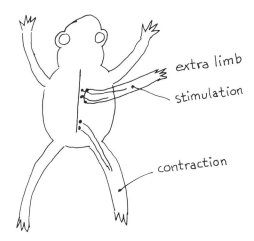

in the leg bud of the tadpole must have told the
nerve fibers that reach it that it is a leg. That
is, some chemical information must be characteristic of leg-like organs. Sperry's idea is that in
the course of development every tissue cell may
acquire a specific set of chemical signals that
inform the arriving nerve of what type of cell it
has reached. Once the nerve cells in the CNS have

been told, that is, once they have been instructed, they will, when the nerve is cut, send fibers to reestablish precise contact with those or similar cells.

One can go even further with experiments on the optic nerve of the frog, which consists of fibers from neurons located in the eye retina. If the optic nerve is cut behind the eye, the frog becomes unable to see. (Testing is done with flying spots on a screen or other devices, recording impulses in the brain cells or just head movements.) After a few weeks the optic nerve regenerates and sight is restored. What happens if the optic nerve is cut, the eye is rotated 180°, and the nerve is allowed to regenerate? When eye-sight returns the response to visual stimulation in that eye is exactly 180° off. This indicates that the nerve fibers have returned to their original connections. Even though the peripheral contacts may originally be responsible for programming the nerve, this programming becomes so well established that by the time the animal reaches adulthood the nerve is forced to reestablish the original connections in the brain.

Nerve fibers from cell bodies in the retina reach specific parts of the brain. One can establish a map correlating specific parts of the retina with sites on the brain. (This was done using goldfish.) If the optic nerve is cut and at the same time a part of the retina is removed, the only nerve fibers that grow back are those from the parts of the retina that remain. But no fibers will reach that part of the brain that corresponds to the lost piece of retina; in fact, that part of the brain degenerates. These observations negate the plausible but naive hypothesis that regenerating fibers are attracted to their previous terminals just because these are now without innervation. There must be at least two signals: one in the retina telling the fibers where to go, and one in the brain telling the wrong axons to stay away. What are these signals?

Nerve growth factor   An old friend of mine, Rita Levi-Montalcini, found out why nerve fibers grow faster in the chick embryo than in the adult. She discovered that some tumors and certain other tissues stimulate nerve growth in the adult and isolated a substance that was responsible for the stimulation. This nerve growth factor is a small protein present

in most animals. In cell cultures as well as <u>in
vivo</u> it stimulates growth of the nerve fibers of
sympathetic neurons, those that innervate the
inner organs of the body. The factor is synthesized in the central nervous system, is distributed
by the circulatory system and, strangely enough,
is excreted in the salivary glands. An antibody
against the factor can inhibit its effect. If
such an antibody is injected into a baby chick or
mouse, the animal fails to develop any sympathetic
nervous system at all: the stimulus for the formation of the sympathetic nervous system has been
destroyed at a critical point in its development.
They are sympathectomized in infancy! Unfortunately these experiments say little as yet about the
specificity of the growth factor, except its pre-
ference for sympathetic cells, or about the control of its production or the mechanism of its
action.

Lecture 36

The nervous system acts in a real sense as the
intermediary between the world and the organism.
It does so by processing stimuli received by
specialized receptors into patterns of nerve impulses transmitted to the central organs. These
impulses may then generate responses, either at a
simple automatic level as in the knee tap reflex
or after further processing, which in man may involve conscious transformations.

Some receptors inform about the external world,
others about internal events within the organism.
The stretch receptor, with which we are already
familiar, is of the second kind. In addition to
receptors in muscles there are four major classes
of sensory receptors in the body: pressure receptors, chemoreceptors, mechanoreceptors, and light
receptors.

Pressure receptors are located in the skin.
They consist of specialized cells surrounding a
sensitive nerve ending that communicates signals
to the CNS. Chemoreceptors are located in the
nose and mouth, but the two classes are organized
differently. The taste receptors in the mouth consist of specialized cells surrounded by nerve end-

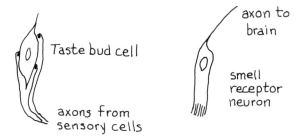

ings, which transmit signals to the brain. The
smell receptors are peripheral neurons, located in
the mucosa of the nose, whose axons proceed directly to the brain. The nerve filaments from the
smell receptors pass through tiny holes in the
base of the skull, between the eyes, and are vulnerable to damage. Many people lose their sense
of smell permanently in car accidents because the
tiny fibers that pass from the nose to the frontal
part of the brain are broken. In general, the
chemical receptor cells undergo changes in permeability to Na^+ in response to an appropriate
stimulus. Influx of Na^+ and depolarization initiates an action potential. But little is known of

Mechanoreceptors

the exact mechanisms involved.

Mechanoreceptors are located, for example, in
the inner ear of mammals, where some of them gather
stimuli of sound and others register changes of
gravity, informing about the orientation of the
body. Mechanoreceptors are known in many animals,
from worms to insects to mammals. They are
built on a generally uniform plan. All mechanoreceptors consist schematically of a cell body
that sits on top of the end of a sensory nerve
fiber. The other end of the cell has cilia,
that is, filaments that extend into the surrounding
liquid, for example in the inner ear canals. At
the base of the cilia there is a basal body consisting of a centriole similar to those seen at
the poles of a mitotic figure:

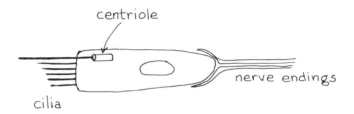

In the acoustic receptor of the inner ear, there
are hundreds of thousands of receptor cells
arranged in a spiral, and covered with a thick
sticky substance. When a sound is received the
covering material oscillates and the resulting
bending of the cilia causes (how??) a depolarization of the membrane of the receptor cell, which
if strong enough generates an action potential in
the nerve fiber and consequently a signal to the
brain. Different mechanoreceptors in the human
ear are sensitive to different wavelengths of
sound within a range of 20 to 20,000 cycles per
second.

Visual receptors

It is somewhat easier to study vision than
hearing. One can easily regulate a visual
stimulus such as a pencil of light or a dark line
in a bright field, and position it precisely in
the visual field or move it around at a controlled
rate. Many experiments of this sort have been
performed on cats, mapping the receptive field of
individual cells, that is, the areas in the visual
field that elicit responses from those cells.
Measurements are made with electrodes implanted in
the retina or in the brain. One can ascertain

whether an electrode has actually entered a cell
by monitoring the potential: if it is in a cell,
the resting potential should register about -70 mV.

Shining light on the retina stimulates the
nerve cells in the illuminated part of the retina.
Fibers from the <u>ganglion</u> cells of the retina of
one eye converge forming an optic nerve behind the
eye. The two optic nerves meet further back at
the optic <u>chiasma</u>. There some of the nerve fibers
cross so that the fibers from the left part of the
left eye and those from the left part of the right
eye proceed together to the left hemisphere, stop-
ping at a region called the left lateral geniculate body. From there, the neurons of the geniculate body send fibers to one area of the left part
of the brain, the optical cortex. Conversely,
fibers from the right part of both eyes converge
toward the right half of the brain. By these arrangements, stimuli in the left part of the visual
field are sensed in the right hemisphere of the
brain and those in the right part of the visual
field are sensed in the left hemisphere. An individual standing at the home plate on a baseball
diamond and looking ahead sees first base with
the left hemisphere and third base with the right
hemisphere.

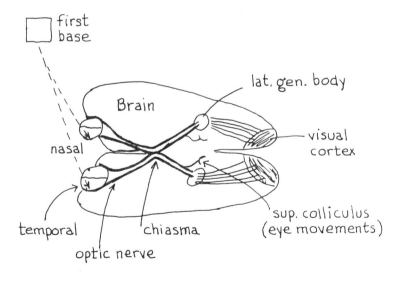

Retina

So much for the connection; now we must explore
the underlying structure. The structure of the
retina itself is complicated. In a section of the
eye one finds, starting from the outside, a tough
protective layer; then a pigmented vascular layer

that stops light from going through; then comes
the retina, itself composed of many layers. The
outermost layer (the most distant from the pupil
of the eye) contains two types of light-sensitive
cells, rods and cones. The cones are scattered
among the rods except for a spot called fovea
near the center, where there are only cones.

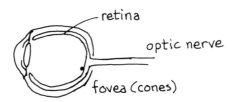

There are three classes of cones, which are responsible for color vision: each class is sensitive to one of three overlapping ranges of wavelengths of light (maxima at 450, 545, 580 nm).

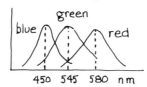

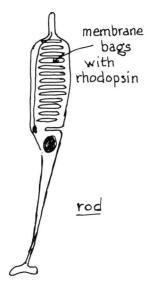

Both rods and cones have an outer segment, loaded
with light sensitive pigment in a stack of flattened membrane bags. The visual excitation results from absorption of light in these bags,
which in turn causes in some way a depolarization
in the outer membrane of the cell. This depolarization generates an action potential that can be
transmitted synaptically to other cells.

How much light is required on the eye to
generate a signal to the brain? Experiments (in
man; oral response, no electrodes!) involving
stimulation of the retina with decreasing amounts
of light indicate that one quantum of light in a
cell is sufficient to generate a one-cell signal,
but that more than one cell (on the order of 4 to
10 cells) must send off a signal for sensation
to be recorded. This requirement for the firing
of several cells at a time in order to generate a
sensation is a reasonable one: it serves to keep
down the "noise level" by preventing stray firings

from a single cell to be interpreted as light signals.

The sensing of light stimuli by the retina is done by a specific pigment which is present in the membrane sacs and can exist in two forms, retinol and retinal, related to vitamin A. Retinol has an alcohol group at the end, retinal is an aldehyde. This would seem a rather small change: but there are other differences. In the retinal form the molecule is bent and is attached to a specific protein called <u>opsin</u>. The combination is called <u>rhodopsin</u>. In response to a light stimulus, retinal is converted to retinol, which has a straight chain configuration and comes off the opsin. The conversion to retinol is the first step in the series of poorly understood reactions leading to the depolarization that causes an impulse to go off. Retinol is reconverted to retinal by an enzyme to restore sensitivity. [An exciting analogy system has been discovered in a bacterium that has patches of membrane made of rhodopsin plus phospholipids. In response to light this membrane becomes energized, so that it can accomplish work including the synthesis of ATP. Maybe the mystery of transduction of light energy into chemical energy will be solved not on eyes but on bacteria!]

Processing of visual signals

The signals from rods and cones must then be transmitted to the brain. The most important point is that these signals are processed several times en route to the brain, so that at each level the signals convey more complex, integrated representations of the visual field rather than just an image of the light patterns on the retina. The first processing steps take place in the retina itself.

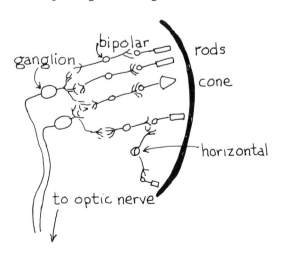

Three layers of cells are involved; in each layer
the cells receive sets of signals from one or
many receptor cells and combine them by positive
and negative reinforcements (stimulatory or inhibitory synapses). The first integration occurs
in the ganglion cells, which send their axons to
the brain. These cells fire spontaneously all
the time at a steady frequency. Their response
to light changes is a change in frequency of
firing: an <u>on</u> response is an increase in the
frequency of firing, an <u>off</u> response a decrease.
Each ganglion cell gets signals from a certain
local territory of the retina, that is, from a
specific area of the visual field. A ganglion
cell can be an <u>on center</u> cell or an <u>off center</u> cell:

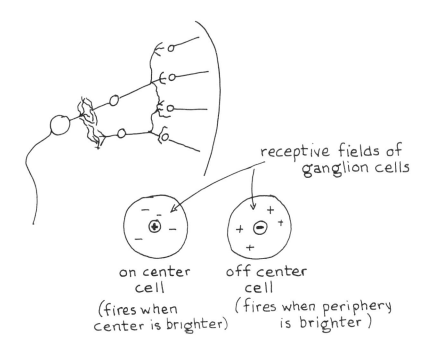

on center   off center
cell   cell
(fires when   (fires when periphery
center is brighter)   is brighter)

That is, some cells fire when the center of their
territory is brighter than the periphery, others
do the reverse. The key point is that each ganglion cell senses a contrast of illumination; the
information is already partly integrated at this
level by a combination of excitatory and inhibitory
stimuli coming to the body and dendrites of the
ganglion cells from the cones, rods, and intermediate neurons of a local retinal region. Diffuse uniform light over the entire territory of a
ganglion cell produces no contrast, hence no signal.

Then the axons proceed to the lateral geniculate bodies and distribute themselves in orderly
fashion among layers of cells. (A few fiber
branches go to another brain area that probably
controls eye movements.) There is at this point
some integration of signals from the two eyes.
Each nerve cell of the geniculate bodies receives
signals from many ganglion cells and gives a frequency response that is again the integrated combination of the signals it gets. The final stages
of processing occur in the visual cortex, where
there are different levels of cells that differ
in the extent of their receptive fields and in
the way they analyze them. There are about 10^5
cells for every square millimeter of cortex surface, arranged in layers and joined mainly by up
and down fibers that make synaptic connections of
many kinds. By-and-large the cortex has a colum-
nar organization: that is, columns of multiply
connected cells are arranged perpendicular to the
surface of the brain, with fewer connections between columns. And each column turns out to be a
functional unit, in which "simple cells" with
different receptive fields but similar field ori-
entations, converge their information on "complex
cells," which integrate the visual signals from
larger fields.

Each simple cell in the visual cortex receives
inputs from several cells in the geniculate body.

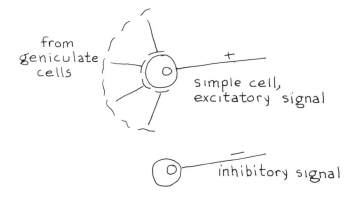

Each cell has a field divided into excitatory and
inhibitory regions; but here the regions are not
center versus surround, as in the ganglion cells,
but parallel strips:

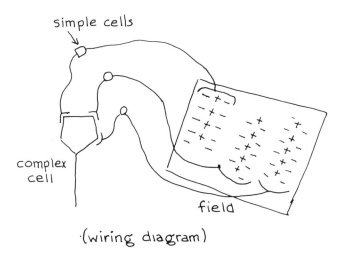

receptive
field of
simple cell

{ − inhibitory region
  + excitatory region

The cell responds to lateral displacements of an object in the visual field. In turn, many simple cells send signals to the complex cells in the same column. These cells analyze a larger field and respond to the width and orientation of a visual object over their field, integrating the information from many simple cells. There are

simple cells

complex cell

field

(wiring diagram)

probably also the so-called hypercomplex cells, which zero in on more precise features of the field: lengths of objects, curves, corners, and angles within and between them: in other words, shape and contour. The final outcome is the visual field as it appears in our minds. Note that the cortical representation is not like a photograph, that is, a point-by-point image. It is a multistage analysis and interpretation of the surrounding reality.

Visual deprivation   What about binocular vision, which makes possible the precise location of objects in depth? In the cat cortex about 80 percent of the cells respond to both eyes, 20 percent to one eye only. But even the binocular cells discriminate: each is dominated by one or the other eye, so that dif-

ferential analysis of the visual field is possible.
The remarkable thing is that the binocular wiring,
or at least its performance, is determined not
only genetically but also developmentally. If one
covers the left eye of a newborn kitten until it
is three months old (visual deprivation) and then
analyzes the cortical responses, one finds that
almost 90 percent of the cells are used for the
right eye only! The connections from the inexperi-
enced eye either have not formed or have become in-
effective, and this situation is irreversible.
The trouble is only at the cortical level: ganglion cells and geniculate cells are normally
connected.

Many experiments with visual deprivation and
artificial cross-eyedness have shown that the
critical thing in properly wiring the cortex is
the eye synergy, that is, the simultaneous activation of the visual field of the two eyes at a
critical period during development. If one eye
does not send signals in that period of time, the
other eye takes over. Presumably, its signals
inhibit the formation or activation of synapses
by fibers from the idle eye.

These visual deprivation phenomena pose sharply
the dilemma of heredity and environment. The
genetic system is poised to create a functional
apparatus provided the stimuli that have served to
select it in the course of evolution are properly
provided in the course of development. If not,
then a new arrangement may be adopted that is more-
or-less suitable to the altered situation. Thus
there is a certain degree of plasticity in the
cortical wiring diagram, and this plasticity is in
fact what makes it possible to create a functional
circuitry in response to normal development. For
example, we now believe that human language, that
unique instrument of the human mind, shares with
the visual system the duality of a genetically
programmed grammatical capacity and an environ-
mentally determined linguistic content. An even
clearer example of this duality of programming is
the imprinting phenomenon: a newly hatched gos-
ling or chick learns to follow whatever walking
animal it comes upon at a critical time. It may
follow its mother, another bird, or an experi-
menter, and from then on continues to do so. The
genes tell it to follow, the environment tells it
whom.

One could go on to talk of behavior, nature-nurture problems, and other aspects of psychobiology in terms of evolution. But here we must stop. Before we do so, however, take a few minutes--or longer--to reflect on what the organization of the visual system implies. From the single cell of the fertilized egg a few thousand genes, directing the structure of as many proteins, construct in a few weeks or months a light field analyzer such as no computer can yet emulate, and what is more, they construct it the same in every individual. Even more fantastic at first seems to be that evolution could have created not only such a perfect instrument as our visual system but also a series of almost equally perfect but anatomically very different ones in insects, mollusks, and other animals. The answer, of course, is that the two processes--development and evolution--are in fact one. Development can be fast, precise, and stereotyped because over the billions of years evolution was fumbling, imprecise, and opportunistic--in other words, existential. Evolution did its best by just keeping life going and is still doing so.

In this book our approach to biology has been through what we may call the two most elementary parts of biology: biochemistry and genetics. We have not dealt with the most formal aspects of evolution, taxonomy or ecology, which properly belong to a more advanced and specialized study of biological phenomena. We have schematized the study of organisms around the concept of the genetic program and its functional expression, that is, the duality between the genes and their environment. The reader will have noticed, however, the existence of another important duality in biology, that between mechanical and phenomenological interpretations. On the one hand, each phenomenon of biology involves functions and structures of organisms. These functions and structures imply biochemical mechanisms that reflect the activity or inactivity of certain genes. This is the mechanical aspect. On the other hand, each phenomenon or group of phenomena has a pattern of its own irrespective of the underlying biochemical mechanisms, and that pattern must be described and understood on its own level.

For example, the circuitry of the reflex arc or of the mammalian visual system has an organization, a biological reality of its own. Knowledge of the mechanisms that bring about cell-cell recognition, synapse formation, and the final wiring will add further dimensions to our understanding of these phenomena, but this is not a prerequisite for a description of the system at the structural and functional level. In biology, phenomenology can stand on its own.

Why is that? The answer is, once more, the existential quality of evolution. What exists is the almost accidental product of a purposeless historical process. A given visual system or a given device for gene regulation is but an exceptional sample of all visual systems or all gene regulatory devices that could have come into being. Diversity in biology is not like a series of displacements along a continuous space. Neither is it like a series of jumps from star to star through an empty space. The organisms that actually exist are like a group of islands that are but the peaks of a submerged mountain chain. The species and organ systems that we currently see are the adaptive peaks of an ever changing profile. It is the diversity and uniqueness of biological types and devices that make them phenomenologically significant.

Enrico Fermi used to say that a miracle is an event with less than 10^{-1} probability of taking place. If so, then every biological object--a species, an organism, an organ--is a miracle. As such, their features may well be approached with the feeling of wonder that miracles demand. The desire to take them apart to see what hides inside should be balanced with the reverence that everything rare and precious deserves.

CHEMICAL BACKGROUND

Matter is composed of atoms and molecules. An atom consists of a positively charged nucleus surrounded by negatively charged electrons. The
electrons occupy specific orbits of characteristic energy around the
nucleus; each orbit can accommodate a fixed maximum number of electrons.
In general, the electrons in the outermost orbit determine chemical behavior: atoms with the same configurations in their outer orbits tend
to have similar chemical properties. Since inner filled orbits are not
directly involved in chemical bonding, they are usually omitted from
chemical representations. Atoms with filled outer electron orbits are
called inert gases and are chemically unreactive; these electronic configurations (filled outer orbit) are particularly stable.

The electronegativity of an atom is a measure of its electron attract-
ing power: an atom which is only one or two electrons short of having a
filled outer orbit is highly electronegative (it tends to attract electrons to fill its outer orbit); an atom with only one or two electrons
in its outer orbit is not very electronegative (it tends to reach inert
gas configuration by losing the few electrons in its outer orbit). The
oxygen atom :Ö: is more electronegative than :Ṅ: or :C:

Molecules are clusters of atoms held together by chemical bonds. In
forming molecules, atoms approach the nearest inert gas configuration by
either (1) gaining or losing electrons (ionic bonds) or (2) sharing
electrons (covalent bonds).

IONIC BONDS. The transfer of electrons from a less electronegative to a
more electronegative atom (or group of atoms) results in the formation

of two charged particles called <u>ions</u>. The electrostatic attraction be-

tween these ions holds them together in an ionic bond. Since the

charges on the ions are spherically distributed, each positive ion can

attract several negative ions and vice versa. The result is the forma-

tion of solid aggregates (crystals) with each ion surrounded by several

of the opposite charge.

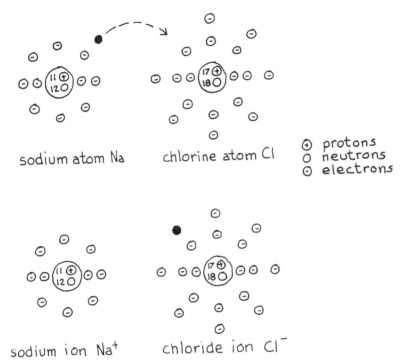

COVALENT BONDS. Atoms of similar electronegativity can approach inert

gas configuration by sharing electrons. The attraction of the two nuclei

for the shared electrons holds the nuclei together. Many atoms can co-

valently share more than one pair of electrons and form multiple bonds.

(In the chemical formulas, a dot represents an electron in the atom's outer orbit; a dash represents a pair of shared electrons. An unshared pair of electrons is called a <u>lone pair</u>. Note that hydrogen can co-valently share no more than two electrons (reaching the helium configura-tion). Elements with up to 7 electrons in the second orbit (for example, C, N, O) can share no more than eight electrons.)

Covalent bonds are shorter and stronger than ionic bonds. Single covalent bonds are longer than double covalent bonds, and double co-valent bonds are longer than triple covalent bonds.

$$-\overset{|}{\underset{|}{C}}-\overset{|}{\underset{|}{C}}- \qquad \diagup\!\!\!C=C\diagdown \qquad -C\equiv C- \qquad \diagdown\!\!\!C-\ddot{O}\!: \qquad \diagup\!\!\!C=\ddot{O}\!:$$

1.54 Å   1.39 Å   1.20 Å   1.42 Å   1.28 Å

Note: The -C=O bond is called a <u>carbonyl</u> bond.

<u>POLAR COVALENT BONDS</u>. When covalent bonds are formed between atoms of differing electronegativity, the bonding electrons are more strongly attracted by the more electronegative atom with the result that this atom bears a slightly negative charge (δ^-) while the less electronega-tive atom bears a slightly positive charge (δ^+). Bonds of this type are said to be <u>polar</u>; they represent an intermediate case between ionic and symmetrical covalent bonds (between like atoms). Polar bonds confer on the molecule a <u>dipole moment</u> and increase affinity for water, also a polar substance.

$$(\delta^+)\quad H \diagdown \atop H \diagup \overset{..}{O}\!:\,(\delta^-) \qquad (\delta^+)H-\overset{(\delta^-)}{\underset{..}{N}}\!\overset{H\,(\delta^+)}{\diagdown H\,(\delta^+)}$$

   Hydrogen bonds are weak interactions which tend to occur between any

electronegative atom (for example, O, N) and a hydrogen atom covalently

bonded to another electronegative atom. In water, for example, hydrogen

bonds are formed due to the attraction between the partial negative

charge on the oxygen of one water molecule and the partial positive

charge on the hydrogen of an adjacent water molecule.

   Ionic compounds such as NaCl dissolve (dissociate into ions) in

water or any polar solvent because the polar molecules of the solvent

form weak bonds with each ion. The presence of ions dissolved in water

confers to the solution the ability to conduct electric current. In

order for a molecule to dissolve in water, it must be able to form elec-

trostatic bonds with water molecules; therefore, it must possess a cer-

tain number of polar or ionized groups.

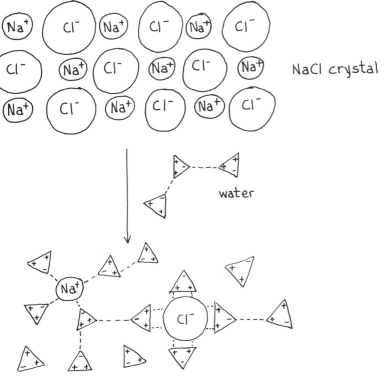

NaCl crystal

water

Na⁺ Cl⁻ in solution

CHEMICAL REACTIONS. LAW OF DEFINITE PROPORTIONS. In the molecules of a given compound, atoms are always combined in the same proportions; the ratio of the weights of the component atoms is also constant. A convenient way to relate numbers of molecules and their weight exists: a mole is that quantity of a substance that contains 6.02×10^{23} molecules. One mole of a compound is equal to the molecular weight of that compound expressed in grams.

	$2H_2$	$+$	O_2	$\rightarrow$	$2H_2O$
number of molecules participating in the reaction	2		1		2
relative mass of reacting substances	4		32		36
molecular weight	2		32		18
moles of substrate or product that disappear or appear	2		1		2

The weight of an atom of H is 1.67×10^{-24} gram. 6.02×10^{23} atoms of hydrogen weigh 1 gram ($1/6.02 \times 10^{23} = 1.67 \times 10^{-24}$). The number 6.02×10^{23} is Avogadro's number, which is also the number of molecules in one mole of any substance.

EQUILIBRIUM CONSTANT. Experiments show that—

1. all chemical reactions are to some extent reversible and can therefore be written $A + B \rightleftharpoons C + D$, and

2. the molar concentrations (expressed as moles/liter) of A, B, C, and D at equilibrium satisfy the equation:

$$K_{eq} = \frac{[C][D]}{[A][B]} .$$

(The square brackets stand for molar concentrations.) K_{eq}, the equilib-

rium constant, is a constant whose value depends only on the temperature

and the identity of the chemical species involved.   For example, for the

reaction--$2Na + Cl_2 \rightarrow 2NaCl$,

$$K_{eq} = \frac{[NaCl]^2}{[Na]^2 [Cl_2]} .$$

Note that if large amounts of A and B are present to begin with, the

reaction goes in the "forward" (left-to-right) direction to reach equi-

librium; if large amounts of C and D are present to start with, the

reaction must go in the "reverse" direction to reach equilibrium.   What

is the effect of continuously removing species C as soon as it is pro-

duced?

ACIDS AND BASES. An acid is a substance that can dissociate in water to

give a conjugate base plus a proton: $AH \rightleftharpoons A^- + H^+$.   A base is a sub-

stance that releases OH^- in water: $XOH \rightleftharpoons X^+ + OH^-$. A^- is a base because

its addition to water causes the release of OH^-: $A^- + H_2O \rightleftharpoons AH + OH^-$.

Thus a base can either be thought of as a molecule that liberates OH^- or

one that picks up H^+ from water.   Any molecule with an unshared pair of

electrons is a base. A^- is an anion (negative charged ion); H^+ is a

cation.

(In dealing with acids and bases, one is most concerned with the

amounts of H^+ or OH^- that are available: the equivalent weight of an

acid is that weight of the substance that furnishes 1 mole of H^+ ions;

the equivalent weight of a base is that weight of the substance that furnishes 1 mole of OH^- (or takes up 1 mole of H^+). Thus one mole of

HCl equals one equivalent weight of HCl but one mole of H_2SO_4 contains

two equivalent weights of $H_2SO_4 = SO_4^{--} + 2H^+$.)

For an acid,

$$AH \rightleftharpoons A^- + H^+,$$

$$K_{eq} = K_a = \frac{[A^-][H^+]}{[AH]} .$$

Taking logarithms, we have

$$\log K_a = \log [H^+] + \log \frac{[A^-]}{[AH]} .$$

Defining $pK_a = - \log K_a$, and $pH = - \log [H^+]$, we obtain

$$pK_a = pH - \log \frac{[A^-]}{[AH]} .$$

pH is a convenient measure of the concentration of H^+ in solution. High

pH means less H^+; low pH means more H^+. Therefore, pK_a is the pH at

which the concentrations of the acid and its conjugate base are equal.

The pK_a value measures the tendency to pick up or give off protons: the

higher the pK_a, the greater the affinity of the conjugate base for pro-

tons. At $pH \ll pK_a$, an acid is practically not dissociated.

If a substance has more than one acid group, each group has its own

pK_a value. For example, the pK_a values for the three acid functions of

phosphoric acid are

$$H_3PO_4 \rightleftharpoons H_2PO_4^- + H^+ , \quad pK_a \approx 0;$$

$$H_2PO_4^- \rightleftharpoons HPO_4^= + H^+ , \quad pK_a = 7.2 ;$$

$HPO_4^= \rightleftharpoons PO_4^{\equiv} + H^+$, $pK_a = 12.3$.

Note: pK_a values can also be used for bases; for example, ammonia in solution:

$$H^+ + NH_3 \rightleftharpoons NH_4^+ \ , \quad K_b = \frac{[NH_4^+]}{[H^+][NH_3]} = \frac{1}{K_a} \ .$$

For weak acids (which are not completely dissociated in water) the pK_a can be measured as the inflection point in the titration curve, in which one measures the pH as a function of the amount of base (NaOH, for example) added to a solution of an acid.

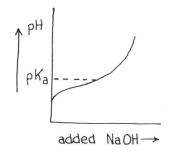

added NaOH⟶

Note in the figure that, near the pK_a value, a solution of a weak acid "resists" changes in pH as more OH^- is added: it acts as a buffer. From the above, pH = pK_a + log $[A^-]/[AH]$. Since K_a, and therefore pK_a, are constant, the pH is determined by the ratio of $[A^-]$ to $[AH]$. In the region of the pK_a both A^- and AH exist in high concentrations. If strong acid is added, it is neutralized by the anion: $A^- + H^+ \rightleftharpoons AH$; if strong base is added, it is neutralized by the acid: $OH^- + AH \rightleftharpoons H_2O + A^-$. Buffers are important because they protect solutions (such as blood) from changes in pH when small amounts of acid or base are added. Examples:

1. What change in pH occurs when 1 ml of 1M (molar) HCl (0.001 liter x

1 mole/liter = 0.001 mole) is added to 1 liter of pure water?

In water some molecules are always dissociated: $H_2O \rightarrow H^+ + OH^-$;

$$K_{eq} = \frac{[H^+][OH^-]}{[H_2O]}$$

The concentration of water is essentially constant (not changed signifi-
cantly by the small amount of ionization), so

$[H_2O] K_{eq} = K_w$,

$K_w = [H^+][OH^-] = 10^{-14}$;

therefore,

$[H^+] = [OH^-] = 10^{-7}$; pH = 7.

Hydrochloric acid is a strong acid and dissociates completely in
water ($HCl \rightarrow H^+ + Cl^-$): 0.001 mole of HCl provides 0.001 mole H^+. The
resulting solution therefore has an H^+ concentration of 10^{-3} M and pH
= 3.

2. What change in pH occurs when the same amount of HCl as above is
added to one liter of a solution containing 1.8 moles of acetate ion
(Ac^-) and 1.0 moles of acetic acid (HAc) per liter?

$HAc \rightarrow H^+ + Ac^-$, $K_a = \frac{[H^+][Ac^-]}{[HAc]} = 1.8 \times 10^{-5}$

The pH of this solution (before adding HCl) is 5.0. Why? What is the
pK_a of acetic acid?

If we add 0.001 mole HCl to this solution, the following reaction
occurs:

$$Ac^- + H^+ \rightarrow HAc,$$

$$K_{eq} = \frac{[HAc]}{[H^+][Ac^-]} = \frac{1}{K_a} = 0.55 \times 10^5$$

Since the equilibrium constant is large, virtually all of the added H^+ reacts with acetate ion to produce acetic acid. The new concentrations are

$$[HAc] = 1.0 + 0.001 = 1.001 \text{ M},$$

$$[Ac^-] = 1.8 - 0.001 = 1.799 \text{ M},$$

and the pH of the solution is virtually unchanged. (Did you understand why?) What would be the effect of adding 0.001 mole of NaOH (a strong base) to the acetate-acetic acid solution above? to pure water?

AMINO ACIDS

Most of the naturally occurring amino acids have four groups bonded to a

central carbon atom (called the α-carbon): (1) a carboxyl ($-\overset{\overset{O}{\|}}{C}-OH$, abbre-

viated -COOH), (2) an amine ($-N{<}^{H}_{H}$, or $-NH_2$), (3) a hydrogen, and (4) an

"R" group which is different for each amino acid (successive carbons of

the R group, if any, are designated β, γ, ...). The carboxyl and amino

groups attached to the α-carbon are called α-carboxyl and α-amino groups.

A carbon atom with four electrons in its outer orbit forms four co-

valent bonds to reach the stable neon configuration. If four single

bonds are formed, they are directed toward the corners of a regular

tetrahedron; this is the energetically most stable configuration since

the mutually repelling electrons in the bonds are the farthest distance

apart. (Why then is H_2O shaped as follows?

Consider the effect of oxygen's lone pairs of electrons.)

In all of the amino acids except glycine (R = H) the α-carbon is

"asymmetric" because it has four different groups attached to it. There

are therefore two possible amino acid configurations, D and L isomers:

(The thin lines represent bonds below the plane; the triangles are bonds

above the plane of the paper. Note: a pair of D and L isomers cannot

be brought to coincide by rotations. They are like a right and a left
hand.)

In proteins, amino acids are always found in the L configuration.
Compounds with asymmetric carbon atoms are "optically active," that is,
they rotate the plane of polarization of polarized light that crosses
their solutions.

In water the carboxyl group readily ionizes:

$$-C\underset{\text{OH}}{\overset{O}{<}} \rightleftharpoons -C\underset{O^-}{\overset{O}{<}} + H^+$$

It is a much stronger acid than, for example, an alcoholic group

$$>C-OH$$

as in serine (see page 376B). The reason is that in the carboxylate
anion the negative charge is shared equally by the two oxygen atoms.
Thus the actual structure is an average of the two possible ones:

$$R-C\underset{O^-}{\overset{O}{<}} \longleftrightarrow R-C\underset{O}{\overset{O^-}{<}}$$

and might best be written:

$$R-C\underset{O}{\overset{O}{<}}(\ominus$$

This <u>delocalization</u> or "spreading out" of charge stabilizes the molecule
(lowers its energy); this phenomenon is called <u>resonance</u> and such a molecule is said to be resonance stabilized.

The nitrogen of the amino group has a lone pair of electrons and acts as a base:

$$-\ddot{N}\!\!\begin{array}{c}\nearrow^H\\\searrow_H\end{array} + H_2O \rightleftharpoons \begin{array}{c}|\\N^{\oplus}\\\diagdown\end{array} + OH^-$$

An amino acid such as alanine therefore has two dissociation reactions:

$$\underset{\text{(acid)}}{NH_3^+-\underset{|}{CH}-COOH} \;\underset{pK_a\atop 2.35}{\overset{-H^+}{\rightleftharpoons}}\; \underset{\text{(acid)}}{NH_3^+-\underset{|}{CH}-COO^-} \;\underset{pK_a\atop 9.87}{\overset{-H^+}{\rightleftharpoons}}\; \underset{\text{(base)}}{NH_2-\underset{|}{CH}-COO^-}$$
$$\qquad CH_3 \qquad\qquad\qquad\qquad CH_3 \qquad\qquad\qquad\qquad CH_3$$

An amino acid whose side chain has another group that can take up or give off a proton has an additional dissociation reaction and pK_a value. Depending on the groups present in the R chain, amino acids are classified as neutral (one carboxyl, one amino group), acidic (two carboxyl, one amino group), or basic (one carboxyl, two amino groups). A chart of the amino acids and their properties is given in the following pages.

CHARGE. The charge on an atom in a molecule is equal to the difference between the positive charge of the nucleus and the sum of the electrons that "belong" to it, that is, all the electrons in its lone pairs plus half the electrons in its shared pairs. The charge on a molecule is equal to the sum of the charges on its composite atoms.

For amino acids the charge changes with the pH:

$$-COOH \rightleftharpoons -COO^- + H^+ \;,$$

$$-NH_2 + H^+ \rightleftharpoons -NH_3^+ .$$

One must sum the charges on all the ionizable groups in the molecule.

When the pH is near the pK_a for one of the dissociable groups, that

group is dissociated on only some of the molecules. At pH = pK_a, a

group is dissociated in 50 percent of the molecules. Thus, at pH 9.87,

the pK_a of the α-amino group of alanine, that amino acid has an average

charge of -0.5 (that is, half the molecules have charge 0, the other

half have charge -1). (Be careful: several amino acids have more than

two dissociable groups.)

   Note: In the $-N\begin{smallmatrix}H\\H\end{smallmatrix}$ configuration (for example, in ammonia NH_3) the

nitrogen (nuclear charge = 5) is uncharged since five electrons "belong"

to it: 1 lone pair plus one electron from each of the three covalently

shared pairs. Nitrogen can gain a hydrogen ion H^+ and lose two electrons

to reach the $\begin{smallmatrix}&N^+\\H&|&H\\&H\end{smallmatrix}$ configuration, in which only four electrons "belong"

to the nitrogen, one from each of the four covalently shared pairs. This

nitrogen form, therefore, has a charge of +1.

PEPTIDE BONDS. In proteins, amino acids are held together by peptide

bonds:

$$H_2N-\overset{\overset{H}{|}}{\underset{\underset{R_1}{|}}{C}}-\overset{\overset{O}{\|}}{C}-N-\overset{\overset{H}{|}}{\underset{\underset{R_2}{|}}{C}}-\overset{\overset{O}{\|}}{C}-N-\overset{\overset{H}{|}}{\underset{\underset{R_3}{|}}{C}}-\overset{\overset{O}{\|}}{C}-$$

The peptide bond is resonance-stabilized by delocalization of the

nitrogen's lone pair of electrons:

$$-\overset{\overset{H}{|}}{\underset{\underset{R_1}{|}}{C}}-\overset{\overset{O}{\|}}{C}-\overset{..}{N}-\overset{\overset{H}{|}}{\underset{\underset{R_2}{|}}{C}}- \rightleftharpoons -\overset{\overset{H}{|}}{\underset{\underset{R_1}{|}}{C}}-\overset{\overset{O^-}{|}}{C}=\overset{+}{N}-\overset{\overset{H}{|}}{\underset{\underset{R_2}{|}}{C}}-$$

Almost no rotation is possible about the peptide bond because of its

partial double-bond character. The six atoms, C-CO-NH-C are in one plane. This places strict limits on the conformations that a protein

chain can assume. Rotation about single bonds is free except for

steric hindrance (bumping together of atoms or groups of atoms).

Table. Amino acids and their properties.

I. ALIPHATIC ("STRAIGHT CHAIN") ACIDS

A. One amino, one carboxyl group.

Formula at pH 7.0	molecular weight	pK_a, 25°C	
		α-carboxyl	α-amino

$H_3\overset{+}{N}.CH.COO^-$
    │
    H
 Glycine (Gly)
	75	2.350	9.778

$H_3\overset{+}{N}.CH.COO^-$
     │
    CH_3
 Alanine (Ala)
	89	2.348	9.867

$H_3\overset{+}{N}.CH.COO^-$
    │
   CH
 $CH_3\quad CH_3$
 Valine (Val)
	117	2.286	9.716

$H_3\overset{+}{N}.CH.COO^-$
     │
    CH_2
    │
   CH
 $CH_3\quad CH_3$
 Leucine (Leu)
	131	2.329	9.752

$H_3\overset{+}{N}.CH.COO^-$
    │
   CH
 $CH_3\quad CH_2$
      │
     CH_3
 Isoleucine (Ileu)
	131	2.318	9.752

$H_3\overset{+}{N}.CH.COO^-$
     │
   CH_2OH
 Serine (Ser)
	105	2.186	9.208

$H_3\overset{+}{N}.CH.COO^-$
    │
   $CHOH$
    │
   CH_3
 Threonine (Thr)
	119	2.088	9.099

B. Sulfur containing monoaminomonocarboxylic acids

	molecular weight	pKa, 25°C		
		α-carboxyl	α-amino	other

$H_3\overset{+}{N}.CH.COO^-$
     $\mid$
     CH_2SH
 	
 Cysteine (Cys)
	121	1.8	8.36	10.53

$H_3\overset{+}{N}.CH.COO^-$
     $\mid$
     CH_2
     $\mid$
     $CH_2S.CH_3$
 	
 Methionine (Met)
	149	2.28	9.21	

C. Dicarboxylic acids and their amides

	molecular weight	α-carboxyl	α-amino	other

$H_3\overset{+}{N}.CH.COO^-$
     $\mid$
     CH_2
     $\mid$
     COO^-
 	
 Aspartic acid (Asp)
	132	1.992	10.004	3.901

$H_3\overset{+}{N}.CH.COO^-$
     $\mid$
     CH_2
     $\mid$
     $CONH_2$
 	
 Asparagine (AspN)
	132	2.18	8.87	

$H_3\overset{+}{N}.CH.COO^-$
     $\mid$
     CH_2
     $\mid$
     CH_2
     $\mid$
     COO^-
 	
 Glutamic acid (Glu)
	146	2.155	9.960	4.324

$H_3\overset{+}{N}.CH.COO^-$
     $\mid$
     CH_2
     $\mid$
     CH_2
     $\mid$
     $CONH_2$
 	
 Glutamine (GluN)
	146	2.17	9.13	

D. Monocarboxylic acids with amine bases

	molecular weight	pK$_a$, 25°C		
		α-carboxyl	α-amino	other
$\overset{+}{H_3N}.CH.COO^-$ $\;$ Lysine (Lys)	147	2.16	9.18	10.79
$\overset{+}{H_3N}.CH.COO^-$ $\;$ Arginine (Arg)	175	1.823	8.991	11.9–13.3
$\overset{+}{H_3N}.CH.COO^-$ $\;$ Histidine (His) at pH 4.0	155	1.82	9.17	6.00

II. AROMATIC AMINO ACIDS

	molecular weight	α-carboxyl	α-amino	other
$\overset{+}{H_3N}.CH.COO^-$ $\;$ Phenylalanine (Phe)	165	1.88	9.13	

II.   AROMATIC AMINO ACIDS (continued)

	molecular weight	pK$_a$, 25°C		
		α-carboxyl	α-amino	other

$\overset{+}{H_3N}.CH.COO^-$
|
CH$_2$
|
...

| 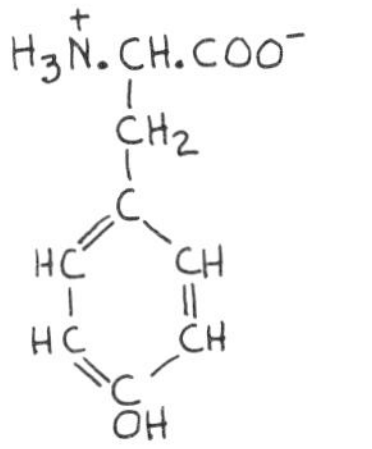 | 181 | 2.34 | 9.11 | 10.13 |

Tyrosine (Tyr)

III.   HETEROCYCLIC AMINO ACIDS

$\overset{+}{H_3N}.CH.COO^-$

| 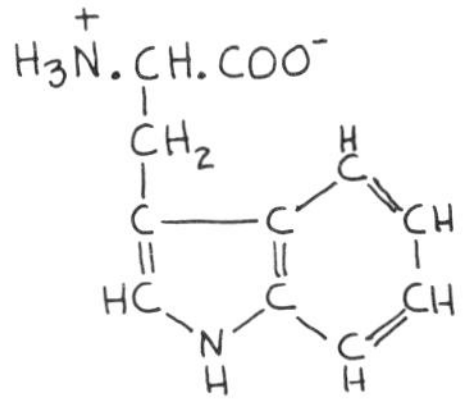 | 204 | 2.39 | 9.39 | |

Tryptophan (Try)

$\overset{+}{H_2N}—CH.COO^-$

| 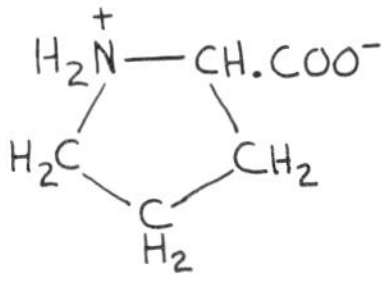 | 115 | 1.952 | 10.643 | |

Proline (Pro)

CARBOHYDRATES (SUGARS)

Monosaccharides (simple sugars) have the empirical formula $(CH_2O)_n$,

$n \geq 3$. The carbon skeleton is unbranched, and most of the carbon atoms

are bonded to a hydroxyl group -OH to form an alcohol. At the remaining

carbon there is a carbonyl oxygen ($-\overset{O}{\overset{\|}{C}}-$). If the carbonyl is at the end

of the chain, the monosaccharide is an aldehyde; if it is at any other

position, the monosaccharide is a ketone.

$$
\begin{array}{cc}
\text{H–C=O} & \text{H} \\
| & | \\
\text{H–C–OH} & \text{H–C–OH} \\
| & | \\
\text{H–C–OH} & \text{C=O} \\
| & | \\
\text{H} & \text{H–C–OH} \\
& | \\
& \text{H}
\end{array}
$$

glyceraldehyde   dihydroxyacetone
(aldehyde)   (ketone)

Aldehydes and ketones are much less stable than carboxylic acids -COOH

because they are less stabilized by resonance. They are chemically more

reactive because of the tendency of the polar carbonyl group $>$C=O to

be attacked by atoms bearing lone pairs of electrons (see page 363B.)

Monosaccharides usually contain one or more asymmetric carbon atoms

and therefore show optical activity. A given chemical formula can repre-

sent several sugars; for example, among the compounds with the formula

$C_6H_{12}O_6$ are

$$
\begin{array}{ccc}
\text{CHO} & \text{CHO} & \text{CHO} \\
| & | & | \\
\text{H–C–OH} & \text{H–C–OH} & \text{HO–C–H} \\
| & | & | \\
\text{HO–C–H} & \text{HO–C–H} & \text{HO–C–H} \\
| & | & | \\
\text{H–C–OH} & \text{HO–C–H} & \text{H–C–OH} \\
| & | & | \\
\text{H–C–OH} & \text{H–C–OH} & \text{H–C–OH} \\
| & | & | \\
\text{CH}_2\text{OH} & \text{CH}_2\text{OH} & \text{CH}_2\text{OH}
\end{array}
$$

glucose   galactose   mannose

The carbonyl group can be attacked by the oxygen atoms of the -OH
groups leading to the formation of various ring forms with 4- or 5- or
6-membered rings. Thus for glucose:

(abbreviated
formula)

(The numbering of carbon atoms is a convention for identifying them.)
In aqueous solutions the chain form is in equilibrium with the ring
form. Five- or six-membered rings are most often formed since they are
sterically the most stable and cause the least amount of strain from
bending and/or stretching of chemical bonds. Various derivatives of
monosaccharides may have the sugar residues in the linear form or one of
the ring forms.

Monosaccharides may be linked together to form polymers: oligo-
saccharides contain from two to ten monosaccharides joined in glycosidic
linkages (removal of H_2O between the C_1 of one sugar and an alcoholic
carbon of another sugar). Polysaccharides contain very long, sometimes
branched chains of monosaccharide units, stored as reserve materials:

Glycogen

5-carbon monosaccharides (ribose and deoxyribose) are parts of nucleic acids. Many proteins, after being completed, get short chains of sugars attached to some of their amino acids.

NUCLEIC ACIDS

The five-carbon sugars ribose and 2-deoxyribose are components of

nucleic acids:

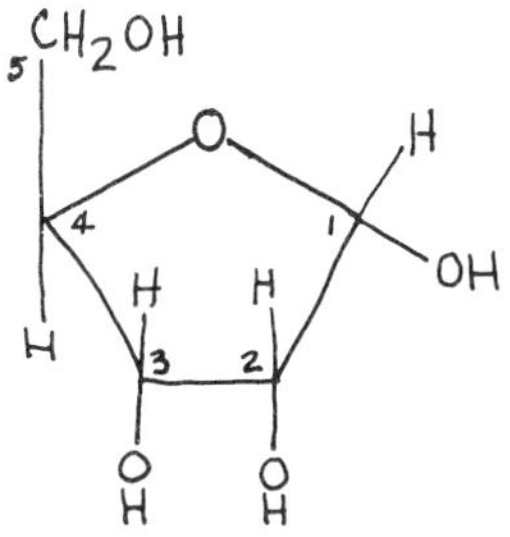

ribose

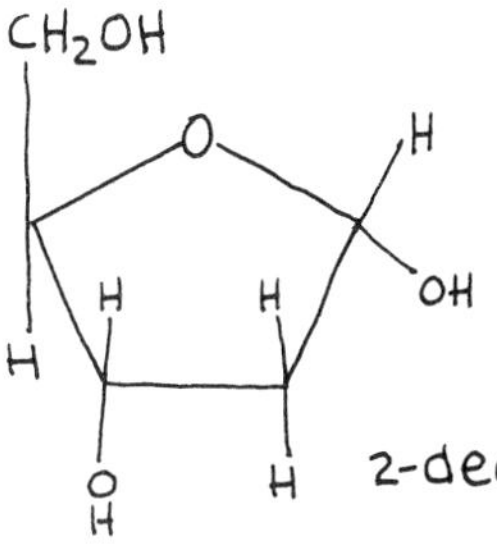

2-deoxyribose

In combination with one of four bases--uracil (in RNA) or thymine (in

DNA), adenine, guanine, and cytosine--these form nucleosides:

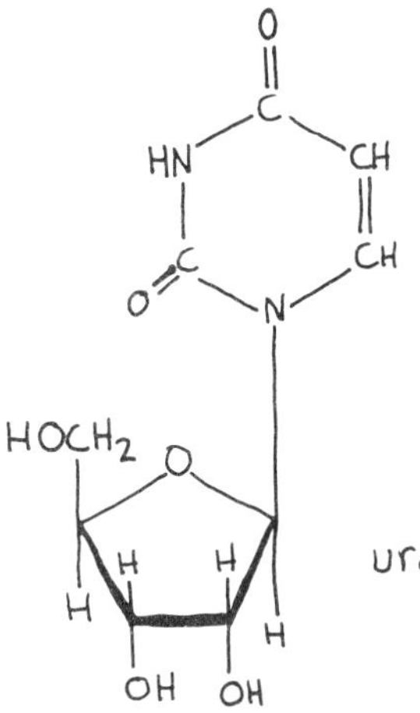

uracil riboside

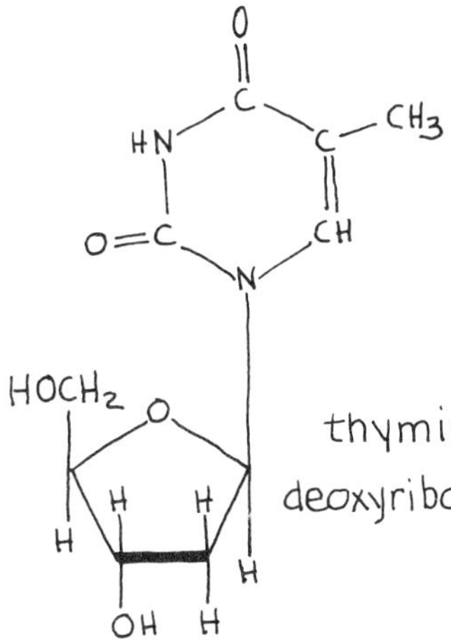

thymine
deoxyriboside

Nucleotides are phosphoric acid esters of nucleosides in which phospho-

ric acid is bound to one of the free hydroxyl groups of the sugar.

Nucleotides containing deoxyribose are called deoxyribonucleotides;

those containing ribose are ribonucleotides.

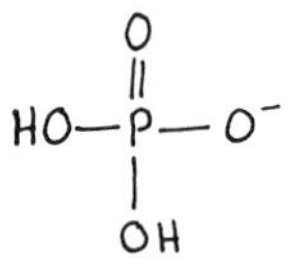

Phosphoric acid
anion

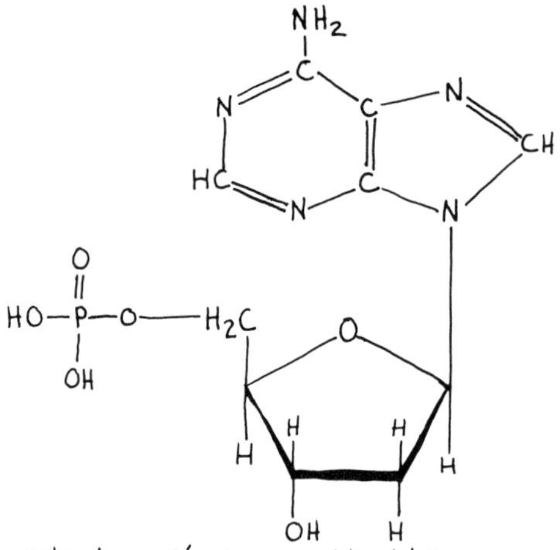

adenine-5'-deoxyribotide
(deoxyadenosine-5'-phosphate)

The 5' indicates that the number 5 carbon of the pentose is bound to

the phosphate group. 2-Deoxyribose indicates that the -OH on position

2 of ribose is replaced by -H. Additional phosphates can be added to

form nucleoside di- or triphosphates.

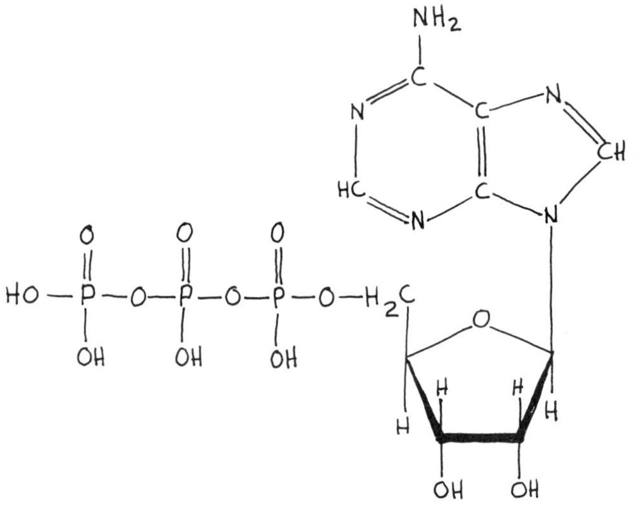

ATP = adenosine triphosphate

Note that the bond between the ribose and phosphate is an <u>ester linkage</u>,

and this is different from the bonds between adjacent phosphates, which

are anhydride linkages. Anhydrides are formed as

the products of elimination of water between acids. They generate acids
when reacting with water.

DNA strands are long chains of deoxyribonucleotides: RNA strands are
long chains of ribonucleotides. The nucleotides are linked by phosphodiester bonds between the 3'-hydroxyl group of the pentose moiety of one
nucleotide and the 5'-hydroxyl group of the pentose of the next nucleotide: (the sugar-phosphate backbone is shaded):

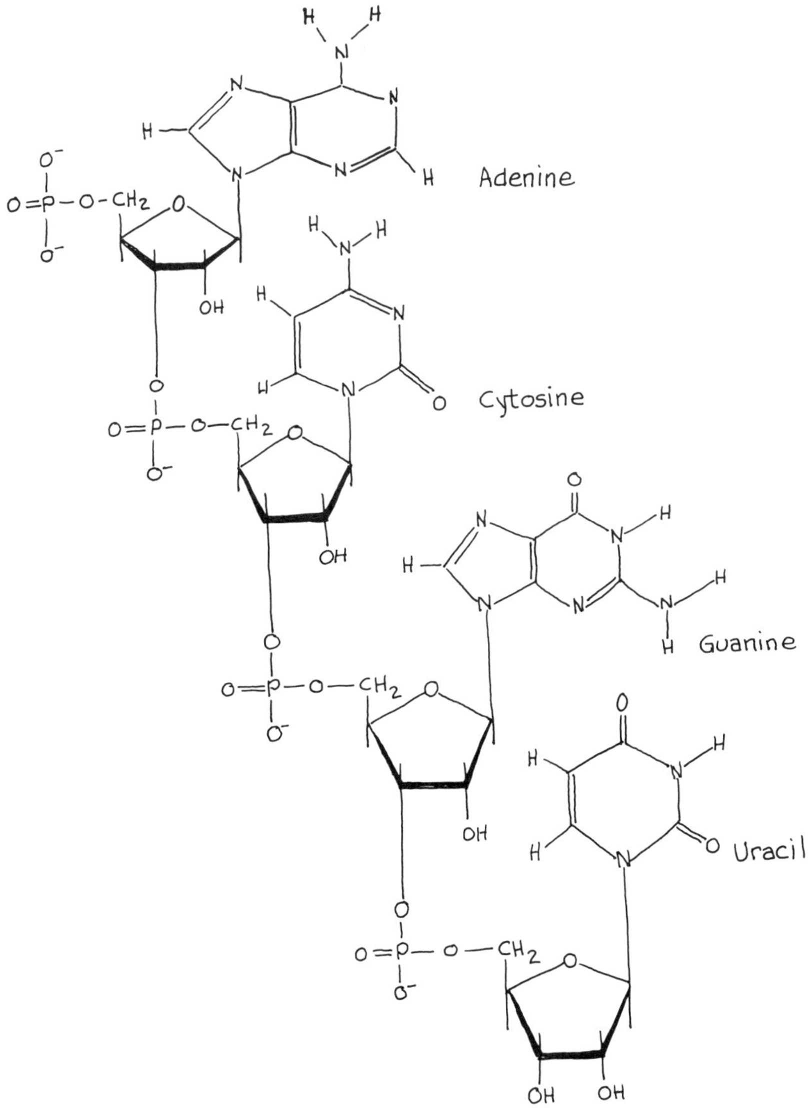

Adenine

Cytosine

Guanine

Uracil

STATISTICAL MECHANICS AND CHEMICAL KINETICS

Not all molecules in a gas or a solution have the same energy. The

frequency distribution function of energies is of the form $N_i = N_0 e^{-\epsilon_i/kT}$

(Boltzmann distribution), where N_0 is the total number of molecules

in any energy state, N_i is the number of molecules in a specific state

whose energy is ϵ_i above the ground state, $k = R/6.02 \times 10^{23}$ is the

molecular gas constant (R = gas constant), and T = absolute temperature.

For a chemical reaction to occur, the energy of the reactant molecules

must be at least as high as the <u>activation energy</u> E_a for that reaction,

that is, the energy difference (per mole) between the energy level of

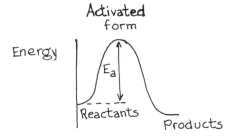

the "activated form" or "transition state" and that of the ground state

of the reactants.

Since the molar fraction with energy $\geq E_a$ (number of moles of the sub-

stance with energy $\geq E_a$/total number of moles) is proportional to $e^{-E_a/RT}$,

the reaction rate is

$$r = A\ e^{-E_a/RT} .$$

$RT \approx 600$ cal/mole at 25°C, represents the average energy of heat motion.

$kT = RT/6.06 \times 10^{23}$ is the average kinetic energy per molecule.

Many reactions involving the <u>breaking of weak bonds</u> (for example,

melting of ice) proceed rapidly without catalysis at normal temperatures
if the ratio of the bond energy to RT is not too high. For example, if
a hydrogen bond has an energy E of 3,000 cal/mole, $E/RT \approx 5$, $e^{-E/RT} \approx e^{-5}$
≈ 0.007; hence the bond is relatively unstable (almost 1 percent of the
molecules have enough kinetic energy to break it in a collision). DNA
strands can be separated by heating at 90°C: $E/RT = 4$.

For reactions involving monoatomic molecules, the reaction rate can
be calculated from kinetic theory, assuming that energy state transitions reflect only the energy of molecular collisions. The rates of
reactions involving multiatomic molecules are often higher than those
calculated from collision theory because, in the impacts between such
molecules, the rotational and vibrational energy of molecules can also
be transferred.

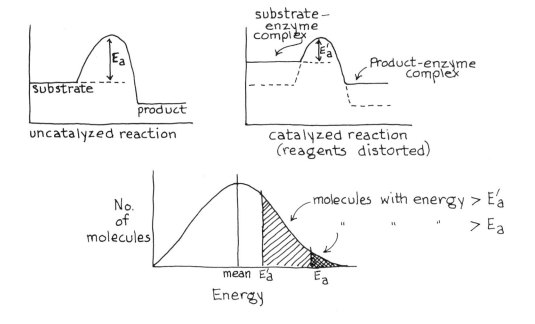

Covalent bonds have high energies, and most reactions involving breaking of covalent bonds do not occur at appreciable rates at room or body temperature. Enzymes accelerate the rates of reactions involving covalent bonds by creating active complexes whose energy level is closer to the activation energy. This is accomplished by constraints produced in the reactant molecules, mainly by weak bonds with the enzyme surface, that is, with the side chains of the amino acids of the enzyme proteins and with coenzymes or prosthetic groups. In this way the participating molecules are forced into a state close to the transition state. In some cases, enzyme-substrate complexes may be stable enough to be isolated as such and may even be held together by covalent bonds.

ENZYME KINETICS

1. Reaction: S → P. (Brackets indicate molar concentrations)

Mass action law: Velocity of product accumulation

$$V = \frac{d[P]}{dt} = - \frac{d[S]}{dt} = k_1[S] ,$$

$$\frac{d[S]}{[S]} = k_1 dt; \quad [S] = [S_0]e^{-k_1 t} .$$

2. Enzyme catalysis by formation of enzyme-substrate complex (E = enzyme).

$$[S + E] \underset{k_2}{\overset{k_1}{\rightleftharpoons}} [ES] \underset{k_4}{\overset{k_3}{\rightleftharpoons}} [P] + [E] \tag{a}$$

If P is removed by other reactions, reaction (a) is practically irreversible $(k_4 \ll k_3)$ and

$$[S] + [E] \underset{k_2}{\overset{k_1}{\rightleftharpoons}} [ES] \overset{k_3}{\rightarrow} [P] + [E] .$$

The velocity of the reaction is

$$V = \frac{d[P]}{dt} = \frac{k_3 \, [S][E_0]}{\dfrac{k_2 + k_3}{k_1} + [S]}$$

where $[E_0] = [E] + [ES] = $ total enzyme concentration.

If we define

$$K_m = \frac{k_2 + k_3}{k_1} ,$$

then

$$V = \frac{k_3 [S][E_0]}{K_m + [S]} = \frac{k_3 [E_0]}{K_m/[S] + 1}$$ (Michaelis-Menten equation).

$k_3 [E_0]$ is called V_{max} and represents the velocity of the reaction for a

given amount of enzyme and for substrate in large excess: $[S] >>> [E]$.

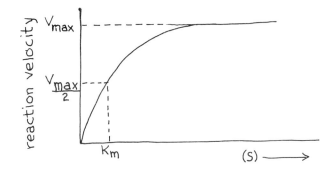

3. Competitive inhibition (I is an inhibitor that competes with S for

E):

$$[I] + [E] \; \underset{k_6}{\overset{k_5}{\rightleftharpoons}} \; [EI] \; ,$$

$$K_I = k_6/k_5 \; ; \; V = \frac{d[P]}{dt} = \frac{V_{max}}{1 + (K_m/[S])(1 + [I]/K_I)}$$

FREE ENERGY CHANGES IN BIOCHEMICAL REACTIONS

1. Definitions.

G = free energy = chemical potential = usable chemical energy

C = concentration of solutes (in moles/liter)

R = gas law constant = 1.987 cal/mole/°C

T = absolute temperature

G = RT ln C + constant   (1)

Free energy change in a reaction

$\Delta G = \Delta E - T\Delta S$ calories/mole

ΔE = change in internal energy of the system

ΔS = change in the entropy of the system

2. Criterion of "feasibility."

Reaction A + B ⇌ C + D proceeds in the direction for which

$\Delta G < 0$.

At equilibrium, $\Delta G = 0$.

3. Relation of ΔG to equilibrium constant.

ΔG depends on two factors:

(a) the intrinsic properties of the reactants;

(b) the concentration ($\approx$ activity) of the reactants.   (Brackets are again

used for concentrations.)

$$\Delta G = \Delta G_0 + RT \ln \frac{[C][D]}{[A][B]}$$ (2)

At equilibrium, $\frac{[C][D]}{[A][B]} = K_{eq}$; $\Delta G = 0$; therefore,

$$\Delta G = \Delta G_0 + RT \ln K_{eq} = 0,$$

$$\Delta G_0 = -RT \ln K_{eq} . \tag{3}$$

Note: K_{eq} is independent of the concentration of the reactants; hence ΔG_0 is a constant dependent only on the intrinsic properties of the reactants. The actual meaning of ΔG_0 is the change of free energy that occurs when a mixture of A, B, C, and D, each with 1 M initially at the start, is allowed to go to equilibrium.

A reaction is <u>endergonic</u> or endothermic (uses up energy) if $\Delta G_0 > 0$, and <u>exergonic</u> or exothermic (releases energy) if $\Delta G_0 < 0$.

All chemical reactions are theoretically reversible. However, the greater the amount of energy released in a reaction, the lower the probability that the product molecules can ever obtain sufficient energy to undergo the reverse reaction. Thus reactions that release large amounts of energy are practically irreversible.

[Note: by convention, the molar concentration of water is taken as unity instead of 55 M since in most reactions with dilute solutions the water concentration remains practically constant.]

4. <u>Role of enzymes</u>.

Although when $\Delta G < 0$ a reaction proceeds spontaneously toward equilibrium, the rate of the reaction may be very slow. <u>Enzymes accelerate the rate of the reactions but do not alter the equilibrium constant</u> (memorize this sentence). Many biochemical reactions proceed extremely slowly in the absence of specific enzymes; hence, even compounds with high group transfer potential (see below) are quite stable in the absence of the proper enzymes.

5. Interpretation and examples.

Problem: What is the ratio of ADP to ATP when the reaction is the hydrolysis of 10^{-4} M ATP in 10^{-2} M phosphate buffer?

Reaction: $ATP + H_2O \rightleftharpoons ADP + H_3PO_4$

(a) Hydrolysis of ATP:

$K_{eq} \approx 10^5$ (measured experimentally)

$$\Delta G_0 = RT \ln K_{eq} = -594 \ln K_{eq} = -1,363 \log_{10} K_{eq} \approx -7,000 \text{ cal/mole} \quad (4)$$

Hence, hydrolysis proceeds spontaneously (but very slowly at body temperature in the absence of the enzyme ATPase).

(b) Specific instance:

$$\Delta G = \Delta G_0 + RT \ln \frac{10^{-2} [ADP]}{[ATP]}$$

$[ATP] + [ADP] = 10^{-4}$ M

Equilibrium is reached when $\Delta G = 0$:

$$\Delta G_0 = -7,000 \text{ cal/mole} = -RT \ln \frac{10^{-2} [ADP]}{[ATP]} = -1,363 \log_{10} \frac{10^{-2} [ADP]}{[ATP]}$$

$$= (-1,363 \times -2)$$

$$- 1,363 \log_{10} \frac{[ADP]}{[ATP]};$$

$$-9,726 = -1,363 \log_{10} \frac{[ADP]}{[ATP]}; \quad \log_{10} \frac{[ADP]}{[ATP]} = \frac{9,726}{1,363} \approx 7 .$$

The equilibrium concentration of ADP will be $\approx 10^7$ times that of ATP; since the total concentration of [ADP + ATP] is 10^{-4} M, the concentration of ATP is $\approx 10^{-11}$ M.

6. <u>Concentration effect and removal of products</u>.

A reaction for which $\Delta G_0 > 0$ proceeds if the concentration of substrates is increased or if the concentration of products is decreased by external sources.

For example: Activation of amino acids by ATP.

ATP + amino acid $\rightleftharpoons$ AMP-amino acid + iPP (inorganic pyrophosphate)

ΔG_0 = 500 cal/mole

K_{eq} = 0.4 (only 40 percent of the amino acid is in the "activated" AMP-aa form).

But iPP + $H_2O \rightleftharpoons$ 2iP (inorganic phosphate)

ΔG_0 = -7,000 cal/mole

$K_{eq} \approx 10^5$

Overall reaction:

ATP + amino acid $\rightleftharpoons$ AMP-amino acid + 2 iP

ΔG_0 = -6,500 cal/mole

$K_{eq} \approx 10^{4.8}$

Almost all the amino acid is activated.

These reactions do not proceed fast unless there are present two enzymes, called amino acid activating enzyme and pyrophosphatase respectively.

[<u>Note</u>: the same principle applies to a number of reactions in the glycolytic pathway, for example, to the reaction

$$CH_2OH-\overset{\overset{\displaystyle O}{\displaystyle \|}}{C}-CH_2O-P \rightleftharpoons CHO-CHOH-CH_2O-P$$

dihydroxyacetone-P glyceraldehyde-3-P   (see page 67).

For the left-to-right reaction, ΔG_0 = +1,800 cal/mole; yet glycolysis

proceeds so that no dihydroxyacetone-P accumulates. Why?]

[Note in section 1: G = RT ln C + constant. This equation repre-

sents the "free energy of dilution." A dilution, say, 1:100, loses

chemical potential because to concentrate back 100-fold would take

energy (heat to evaporate, etc.) This equation also says that when a

cell concentrates a substance from the external medium by means of a

chemical pump it must use energy: "active transport against a concen-

tration gradient." The same is true for excretion against a gradient.]

7. <u>Group-transfer potential and high-energy bonds</u>. (Learn this well!)

Certain compounds have a high "group-transfer potential," that is, the

reactions by which they donate a certain group to other atoms or mole-

cules have high negative ΔG_0. For example, ATP has a high phosphate-

transfer potential, not only to water (as in ATP hydrolysis, section 5

above) but to many other compounds. Compounds with high group-transfer

potential are often called "high-energy compounds," and the bond that

binds the transferred group to the rest of the molecule is called a

"high-energy bond" (indicated by $\sim$). For example, ATP has two such $\sim$

bonds:

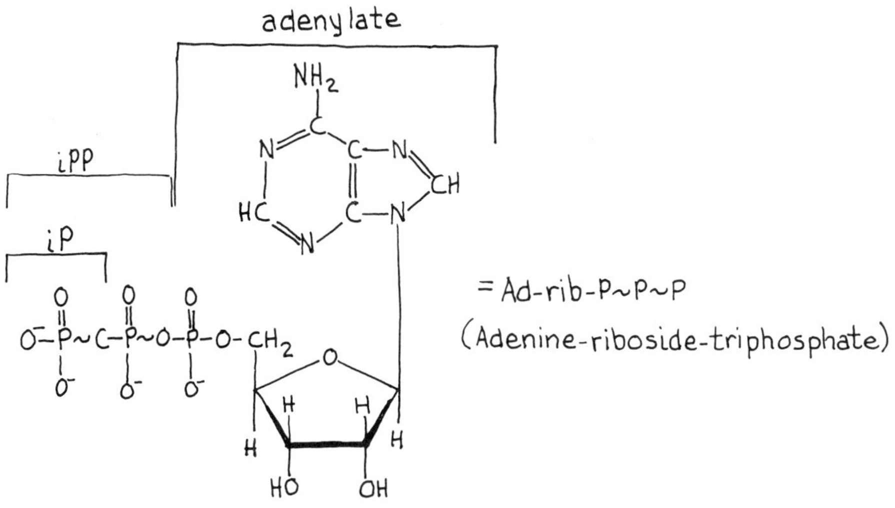

adenylate

$= Ad\text{-}rib\text{-}P{\sim}P{\sim}P$

(Adenine-riboside-triphosphate)

and therefore can readily transfer either a phosphate iP or a pyrophosphate iPP or an adenylate AMP group. There are many such "high-energy"
bonds in compounds of biological importance: acetyl phosphate, creatine
phosphate, amino acyl transfer RNA, etc.

[Note: the expression "high-energy bond" may cause confusion. The
free energy of a group-transfer reaction is not energy derived from
breaking a bond, but is the difference in free energy between the products and the substrates. To break a covalent bond would require an
amount of energy of the order of 100,000 cal/mole.]

The high group transfer potential of compounds such as ATP can be
used in various ways to promote chemical reactions. One such way is by
the "activation" of substrates, as already seen for amino acids. Glucose, for example, cannot be utilized as such, but there are several
pathways which can utilize glucose-6-P:

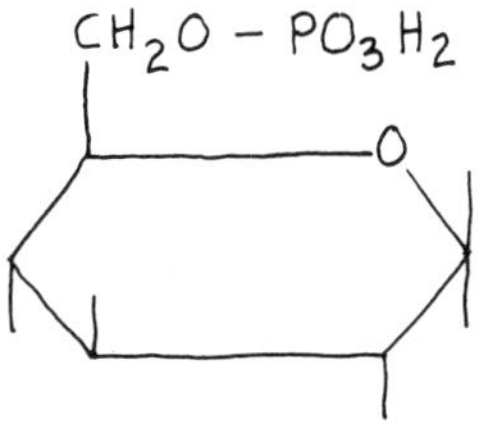

$CH_2O - PO_3H_2$

The reaction of glucose with H_3PO_4 to form glucose-6-P has ΔG_0 = +3000 cal/mole and does not proceed at appreciable rate. Glucose-6-P can easily be formed (in the presence of the proper enzyme) by transfer of a phosphate group from ATP.

glucose + ATP $\rightleftharpoons$ glucose-6-P + ADP; ΔG_0 = -3,400 cal/mole

[Note: while ATP can be written Ad-rib-P$\sim$P$\sim$P, glucose-6-P has no $\sim$P group. The ΔG_0 for the hydrolysis of glucose-6-P is -3,000 cal/mole, not enough for high group transfer potential [must be -(>5,000)cal/mole]. However, glucose-6-P is a chemically more reactive compound than glucose, hence its formation can be considered as an "activation" of glucose, which can thereby enter other reactions.]

ATP is only one of many high group transfer potential substances. For example, NADH (see page 402B) is a high electron-transfer potential substance which can donate electrons to (that is, reduce) many other substances, itself becoming oxidized to NAD.

8. Reactions coupled on an enzyme surface.

In some instances the energy released in a reaction is used to bring about some additional reaction right on the active surface of the enzyme. These enzyme-coupled reactions are very important. For example:

glyceraldehyde-3-P + NAD + iP $\rightleftharpoons$ 1,3 diphosphoglyceric acid + NADH   (1)

(see page 67)

   glutamic acid   glutamine

$$COO^- -CH-CH_2-CH_2-COOH + NH_3 + ATP \rightleftharpoons COO^- -CH-CH_2-CH_2-CONH_2 + ADP + H_3PO_4$$
$$\overset{|}{\underset{NH_3^+}{}} \qquad\qquad\qquad\qquad\qquad\qquad \overset{|}{\underset{NH_3^+}{}}$$

   (2)

9. Supplement for the mechanism-minded student.

What structural features of ATP give it such a large phosphate-transfer

potential?

(a) At pH 7.0 ATP molecules have on the average approximately 4 closely

spaced negative charges which strongly repel each other. When the

terminal phosphate or pyrophosphate group is lost, some of the electrical

stress is removed. The resulting products are anions and have little

tendency to approach each other again because of charge repulsion.

(b) ATP is less resonance-stabilized than the products resulting from

loss of a phosphate or pyrophosphate (and therefore more reactive). The

loss of resonance stabilization in the anhydride linkages between the

terminal phosphates of ATP results from the fact that the lone pairs of

electrons on the oxygen atoms that link the two phosphorus atoms cannot

satisfy the demands of both phosphorus atoms simultaneously. This situ-

ation of "competing resonance" is relieved by the loss of a phosphate or

pyrophosphate, for example:

$$\text{ATP} + \text{H}_2\text{O} \longrightarrow \text{Ad-rib-O} - \overset{\overset{\text{O}}{\|}}{\underset{\underset{\text{O}^-}{|}}{\text{P}}} - \text{O} - \overset{\overset{\text{O}}{\|}}{\underset{\underset{\text{O}^-}{|}}{\text{P}}} - \text{OH} + \text{HO} - \overset{\overset{\text{O}}{\|}}{\underset{\underset{\text{O}^-}{|}}{\text{P}}} - \text{O}^-$$

(1) ADP has approximately three closely spaced negative charges, there-

    fore there is less repulsion than in ATP.

(2) The free phosphate is resonance stabilized:

$$\text{HO} - \overset{\overset{\text{O}}{\|}}{\underset{\underset{\text{O}^-}{|}}{\text{P}}} - \text{O}^- \longleftrightarrow \text{HO} - \overset{\overset{\text{O}^-}{|}}{\underset{\underset{\text{O}^-}{|}}{\text{P}}} = \text{O} \longleftrightarrow \text{HO} - \overset{\overset{\text{O}^-}{|}}{\underset{\underset{\text{O}}{\|}}{\text{P}}} - \text{O}^-$$

and there is less competing resonance in ADP than ATP (only one oxygen

linking phosphorus atoms in ADP).

ACIDS, ALDEHYDES, AND OXIDATION-REDUCTION REACTIONS

Conversion of an aldehyde X-CHO to the corresponding acid X-COOH involves the removal of two electrons and of two protons:

$$X \overset{\overset{\displaystyle :\overset{..}{\underset{..}{O}}:}{}}{:C:H} + H \overset{..}{\underset{..}{:O:}} H \xrightarrow[-2e^-]{-2H^+} X \overset{\overset{\displaystyle :\overset{..}{\underset{..}{O}}:}{}}{:C:} \overset{..}{\underset{..}{O:}} H \quad electrons$$

12   8   18

Aldehydes, as discussed earlier, are much more reactive than acids. The
carbonyl group -C=O has its electrons pulled way out toward the O atom,
which is more electronegative $:\overset{..}{\underset{..}{C}}::\vec{O}$. Therefore the carbonyl C atom
tends to be "attacked" by reagents that are relatively electron rich
(nucleophiles).

Conversion of an aldehyde to an acid is an oxidation, that is, electrons are removed. Addition of electrons is a reduction. An acid can
be reduced to an aldehyde.

Oxidation of an aldehyde to an acid releases lots of energy. It is,
therefore, irreversible unless energy is introduced from the outside.
All reductions of acids to aldehydes require an external input of
energy besides an input of 2 electrons.

Oxidations release energy.

In the glycolysis pathway (page 67) one oxidation reaction (step
number 6) is sufficient to release enough energy to generate two ATP
molecules from ADP + iP, and also to produce one NADH molecule which
still has a relatively high donor potential for 2 electrons.

Note that the oxidation of a compound, for example X-CHO → X-COOH,
is always coupled with the reduction of another compound, for example

$NAD^+ \rightarrow NADH$, since the electrons must be passed on to some other atom.

The redox potential measures the tendency of a compound to donate electrons: $A \rightarrow A^+ + e^-$. In biological systems electrons are most often transferred in pairs. By convention the values for redox potentials E_0' found in tables are for the reverse reaction, $A^+ + e^- \rightarrow A$. Implicit in this equation is the presence of an electron acceptor. Thus a more complete representation is

$$
\begin{aligned}
A &\rightleftharpoons A^+ + e^- \\
B^+ + e^- &\rightleftharpoons B \\
\hline
A + B^+ &\rightleftharpoons A^+ + B
\end{aligned}
$$

The redox potential for the overall reaction is determined by summing the redox potentials for the half reactions:

$$A^+ + e^- \rightleftharpoons A \ ; \ E_a$$
$$B^+ + e^- \rightleftharpoons B \ ; \ E_b$$

Therefore, for the reaction $A + B^+ \rightleftharpoons A^+ + B$, the redox potential (change in electron-donor potential) $\Delta E_0'$ equals $E_b - E_a$. Transfer of electrons between compounds generally involves, like any other reactions, a change ΔG in free energy. $\Delta E_0'$ is proportional to $-\Delta G_0'$. A reaction proceeds in the direction for which $\Delta E_0' > 0$.

The redox potentials are measured by the currents generated in an appropriate electrochemical system. $\Delta E = 1$ volt corresponds to 23 kcal of ΔG for every mole of electrons transferred.

Cells have a number of redox compounds that serve as donors and acceptors of electrons in the course of metabolism. Since chemical energy is

made available in oxidation of compounds and since synthesis of complex
molecules often requires reduction of compounds, the redox intermediates
are extremely important cell constituents.

$NAD^+ + 2\ e^- + H^+ \rightleftharpoons NADH$ is a typical and important example of reversible oxidation involved in many pathways, including glycolysis.   Note
the electron changes at the N atom of nicotinamide.

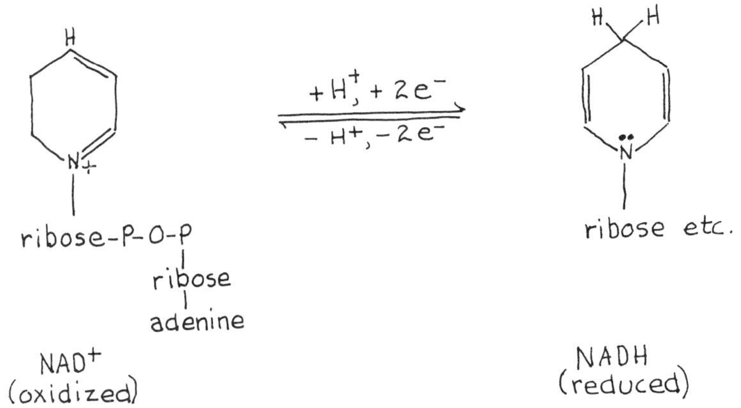

$$\xrightarrow[\ -H^+,\,-2e^-\ ]{\ +H^+,\,+2e^-\ }$$

ribose-P-O-P
        ribose
        adenine

NAD⁺
(oxidized)

ribose etc.

NADH
(reduced)

Nicotinamide adenine dinucleotide

Topics for
discussion

1. What is the relation of the doubling time of a
bacterial culture to its growth rate?

2. What does a bacterium such as E. coli require
for growth, and why? How does it differ from a
mammalian cell line (for example, the human HeLa
cell line)?

3. What distinguishes procaryotes from eucaryotes?
What are organelles?

4. In a given medium, bacteria of strain A have a
doubling time of 30 minutes; bacteria of strain B
have a doubling time of 60 minutes. If you start
a mixed culture with N cells of A and 16N cells
of B, how long will it take for the number of A
cells to catch up with the number of B cells?

5. What is a buffer? Why does a solution of glycine serve to buffer solutions around pH 2.3 - 2.5?

6. What is the equilibrium constant K of a chemical
reaction (for example, $A + B \rightleftharpoons C + D$)?

7. What would be the charge of the dipeptide L-
alanyl-L-alanine at pH 6? for the dipeptide L-
alanyl-α-L-glutamic acid at the same pH?

8. What would be the chemical formula of the amino
acid histidine at pH 9.5? at pH 7.5? at pH 11.0?

9. The pH of human blood serum is kept around 7.
What is the concentration of H^+? A major buffering substance in blood is hemoglobin, a protein in
red blood cells which has about 8% histidine. Can
you see why this is important?

Topics for
discussion

1. What properties of amino acids are responsible
for their separation in electrophoresis and paper
chromatography?
How would you separate the following pairs of
amino acids?
a. glycine and cysteine
b. lysine and alanine

2. How is the three-dimensional shape of proteins
influenced by
a. hydrogen bonds?
b. hydrophobic bonds?
c. disulfide bonds?

3. The enzyme insulin is first synthesized as an
inactive precursor (pro-insulin) containing 81
amino acids. Later, a section of 30 amino acids
is enzymatically removed from the middle of the
pro-insulin molecule. The two remaining end
chains, which are associated by two disulfide
bonds, form the active enzyme insulin. Can you
think of reasons why pro-insulin is inactive?

4. The enzyme trypsin splits peptide bonds on the
carboxyl side of lysine and arginine. What does
this tell you about the specificity of substrate
recognition by trypsin?

5. Treatment of a hexapeptide with trypsin produces
the following peptides and amino acids:

arg + ala-ser + leu-gly-lys.

What are the possible sequences of the amino acids
in this hexapeptide?

6. When the nucleoside 3'-deoxyadenosine is added
to a culture of mammalian cells, the nucleoside
is incorporated into RNA molecules and then further RNA synthesis is terminated.
a. Why?
b. What would be the effect of adding the corresponding nucleotide?

7. Glycogen is a branched polysaccharide consisting of glucose units linked together by removing a
molecule of water. Glycogen consists of series of
glucose joined by 1,4 linkages; branches start
with 1,6 linkages. Do you expect that a single
enzyme can synthesize all of the glycogen? Why or
why not?

Written problem   Draw a curve of the overall charge of the tripep-
                         tide leucyl-lysyl-serine as a function of pH.
                         Indicate the pH values at which the tripeptide can
                         serve as a buffer.

Topics for
discussion

1. The first reaction in the glycolytic pathway is:

$$\text{glucose + ATP} \quad \overset{\text{glucokinase}}{\rightleftharpoons} \quad \text{glucose-6-P + ADP;}$$

$\Delta G_0' = -3.42$ kcal/mole.

a. Which way does the reaction proceed in the absence of the enzyme?
b. Calculate the equilibrium constant of the reaction. Does the presence or absence of the enzyme affect the equilibrium constant?
c. Calculate the ratio of glucose-6-P/glucose at equilibrium, if we start with equal molar amounts of glucose and ATP.

2. Increasing the temperature by 10°C increases the rate of the above reaction by about a factor of 2. Why?
What does this tell you about the activation energy E_a of this reaction?
Calculate an approximate E_a value.
Note that the activation energy E_a is <u>not</u> the $\Delta G_0'$ of a reaction. What is the relation between the two?

3. What is the difference between competitive and noncompetitive inhibition of enzyme activity? How could the two types of inhibition be distinguished experimentally? (Think in terms of V_{max} and K_m.)

4. Reactions 8 and 9 of the glycolytic pathway both have positive $\Delta G_0'$ values:

3-P-glyceric acid $\rightleftharpoons$ 2-P-glyceric acid

$\Delta G_0' = +1.06$ kcal/mole

2-P-glyceric acid $\rightleftharpoons$ phosphoenol-pyruvic acid

$\Delta G_0' = +0.44$ kcal/mole

Yet, during glycolysis both reactions proceed in the left-to-right direction. Why?

5. What is the difference between the reactions

a. ATP + CH_3COOH (acetic acid) $\rightleftharpoons$ $CH_3CO\sim O-\overset{\overset{O}{\|}}{\underset{\underset{OH}{|}}{P}}-O^-$
   (acetylphosphate) + ADP

and

b. ATP + glucose $\rightleftharpoons$ glucose-6-P + ADP?

Which one has the lower (more negative) $\Delta G_0'$ and why?

Topics for
discussion

1. Verify that the reactions of glycolysis plus those of electron transport and those of the Krebs cycle together produce 40 moles of ATP for every mole of glucose converted to CO_2 and H_2O. Why is the <u>net</u> gain 38 moles of ATP per mole of glucose?

2. The aerobic oxidation of $NADH_2$ results in the production of 3 ATP while the oxidation of $FADH_2$ results in the production of only 2 ATP. Why? The inhibitor amytal blocks the transfer of electrons from FP_1 to Coenzyme Q in the electron transport chain. What effect would amytal have on the oxidation of $NADH_2$ by a suspension of mitochondria? of succinate?

3. Based on the principles of glycolysis, postulate a scheme for the aerobic metabolism of fructose, the sugar present in grape juice.

4. Many biological processes are dependent on the conversion of ATP to ADP. What is the difference in the use of ATP in the following reactions?

a. glutamic acid $+NH_3$ + ATP $\rightleftharpoons$ glutamine

   + ADP + H_3PO_4

b. glucose + ATP $\rightleftharpoons$ glucose-6-P + ADP

Written problem

A suspension of minced muscle is given glucose labeled with C^{14} in the number 1 carbon.

a. Where does the labeled carbon end up when the preparation is kept in the absence of O_2? in the presence of O_2?

b. Where will the labeled carbon atom be found after one turn of the Krebs cycle, when the suspension is kept in the presence of O_2? How can you test this experimentally?

Topics for
discussion

1. In an in vitro DNA synthesizing system containing DNA polymerase and template DNA, the addition
of radioactive dTTP (labeled in the thymine) results in a very small amount of radioactively
labeled acid-insoluble product. If you add unlabeled dATP, dGTP, and dCTP, the amount of radioactive acid-insoluble product is greatly increased.
Why?

2. Many bacteria can convert certain amino acids
into Krebs cycle intermediates and, from these,
carry out the synthesis of glucose and other car-
bohydrates.

a. Why is this important for a bacterium growing
with the amino acid glutamic acid as its sole
carbon and energy source?
b. What do you expect E. coli must do in order to
use glutamic acid as an energy source? or to use
alanine?

3. The following genes are found to double in
number at the given times after initiation of
synchronous DNA synthesis in a culture of E. coli:

time	gene
1 min	C
5	A
8	G
10	D
15	E
25	B
35	F

What are the possible sequences of these genes in
the E. coli genome?

Written problem

A short DNA chain has the following sequence:

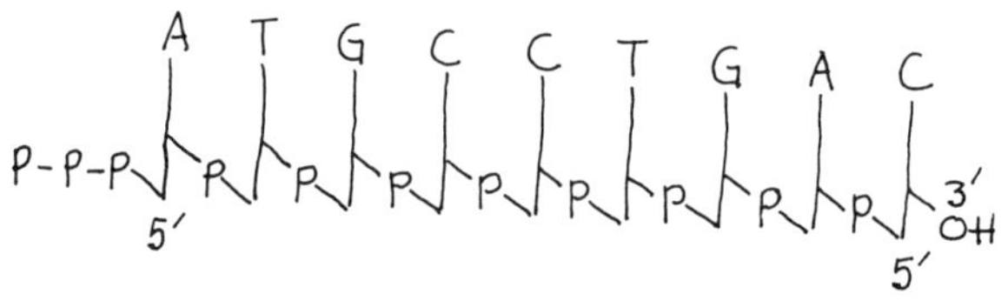

What is the complementary DNA sequence? the complementary RNA sequence? Indicate the polarity
and the direction of synthesis of the complementary sequences.

^{32}P-labeled GTP can be used in the <u>in vitro</u> synthesis of the complementary RNA sequence. Will you find ^{32}P incorporated when the following GTP isotopes are used?

G-$\overset{*}{\text{P}}$-P-P   G-P-$\overset{*}{\text{P}}$-P   G-P-P-$\overset{*}{\text{P}}$

(*P indicates ^{32}P)

Indicate the position of ^{32}P when present in the complementary RNA sequence.

Topics for
discussion

1. The genome of E. coli contains 2.5×10^9 daltons of DNA. (1 base ~ 330 daltons, 1 base pair ~ 660 daltons.) Assume that the average polypeptide chain of E. coli contains 300 amino acids.

a. What is the maximum number of polypeptide chains that E. coli can make?
b. It has been estimated that E. coli contains 2,500 polypeptide chains. As seen from your answer to part (a), there is a difference in the maximum number of polypeptide chains and the estimated number. What could be the function(s) of the extra, nonprotein encoding DNA?

2. In an in vitro system, the A strand of the following DNA sequence is transcribed by RNA polymerase and the resultant mRNA is translated into protein:

5' C T A A C C T G A G C T G G G C A T ... 3'
A strand

3' G A T T G G A C T C G A C C C G T A ... 5'
B strand

a. What is the 5' → 3' sequence of the corresponding segment of mRNA?
b. What is the amino acid sequence coded for by the mRNA?
c. Does this mRNA contain any translation signals?

3. a. How would you determine if genes M and N of Drosophila were linked (on the same chromosome) or unlinked?
b. What would be the order of three linked genes L, M, N if the recombination frequencies were as follows:

$\overline{LM}$ = 2 percent; $\overline{MN}$ = 3 percent; $\overline{LN}$ = 1 percent.

Written problems

1. In the common houseplant Coleus, the genes for leaf margin type and for veination are on separate chromosomes.
Irregular veination (I) is dominant to regular veination (i).
Deeply toothed leaf margin (D) is dominant to shallow toothed margin (d).

a. Determine both the genotype and the phenotype of the P (parental), F_1 (first generation) and F_2 (second generation) plants obtained from the cross:

DD, II   x   dd, ii

(Remember that the F_2 generation comes from inter-breeding the plants of the F_1 generation among themselves.)

b. By what cross can you confirm the predicted F_1 genotype?

2. In tomatoes, the following genes occur on chromosome 2:
+ Tall plant   versus d dwarf plant
+ Green plant   versus l light green plant
+ Smooth fruit versus h hairy fruit
Results of the cross + + + / d l h x d l h / d l h
were:
+ l +   32
+ + + 430
d + h   38
+ + h   4
+ l h   18
d l +   6
d l h 460
d + +   12

a. Which groups represent double crossovers?

b. What is the correct gene sequence?

c. What are the distances, in map units between the first and second genes? the second and third genes? the first and third genes?

d. Do the values fit the theoretical mapping function?

Topics for discussion

1. What is the difference between F^-, F^+ and Hfr strains of E. coli?
How can you distinguish between them?
What is a partial diploid?

2. Distinguish between complementation and recombination.

3. Three mutant strains of E. coli require tryptophan for growth (that is, they are trp^-). The synthesis of tryptophan involves four specific enzymes.

a. How would you determine experimentally if the three trp^- mutations are in the same gene or in different genes?
b. If you find that the three mutations are in the same gene, how would you find out what the order of the mutations is?

4. You isolate a set of seven rII mutants of phage T4 (all rII mutants make large plaques and prevent phage growth on E. coli K(λ) but not on E. coli B).

a. By what test can you verify which of the mutants are in gene rIIA and which are in gene rIIB? Understand the actual details (which bacteria you use, what you test and how).

Written problem

Two strains of E. coli A and B growing in a simple culture medium without histidine synthesize the enzymes of the histidine biosynthetic pathway, of which there are 10.

a. When histidine is added to the culture medium, strain A quickly stops making all 10 of the histidine synthetic enzymes. Explain.

b. Mutant strain B continues to make the histidine synthetic enzymes after the addition of histidine. What are the possible explanations for this result? What test can you use to distinguish between the possibilities?

c. Two other mutants C and D cannot make histidine. You isolate the 10 enzymes and find that, in strain C one of the enzymes (enzyme 3) has an arginine residue in position 30, where there is normally a serine residue. Suggest an explanation for this mutation.

d. In strain D enzyme 3 is missing and, instead, you find a shorter polypeptide which is 29 amino acids in length. Suggest an explanation for this mutation.

Topics for
discussion

1. A carrier of translocation Down's syndrome has
the following chromosomes:

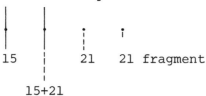

15   21   21 fragment

15+21

Deduce the structure that would be formed when
these chromosomes pair in the prophase of meiosis,
on the assumption that a region of a chromosome
carrying a certain set of information always
manages to come into juxtaposition with the homologous region carrying the same information. What
would be the consequences of a crossover event
between two chromosome 21-regions?

2. In man, the genes for color-blindness and hemophilia are both on the X chromosome with about 10
percent crossing-over. A pedigree involving both
traits looks like this:

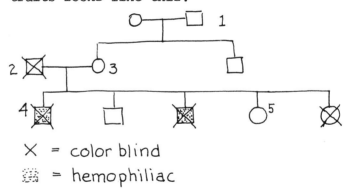

X = color blind
⠿ = hemophiliac

a. Is the allele that produces hemophilia dominant
or recessive?
b. Is the allele that produces color blindness
dominant or recessive?
c. The individuals numbered 1, 2, 3, 4 are among
those in the pedigree whose genotype can be determined with respect to the genes for hemophilia
(H,h) and color-blindness (B,b). What are they?
d. What is the probability that a son of the woman
marked 5 will be color blind? What is the probability that her son will be hemophiliac?

3. On the basis of differing electrophoretic mo-
bilities, two types (A and B) of the enzyme glu-
cose-6-phosphate dehydrogenase (G-6-P-D) can be

distinguished. Individual clones of cultured
cells from a female heterozygous for the alleles
determining A and B respectively show one type of
G-6-P-D or the other, but not both. What is the
implication of this observation?

4. Assume that the pedigree presented below is
straightforward, with no complications such as
mutation or illegitimacy. Trait W, found in
individuals represented by the black symbols, is
rare in the population at large. Which of the
following patterns of transmission for W are consistent with and which are excluded by the pedigree?

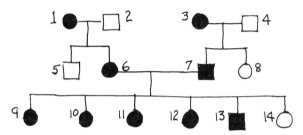

a. autosomal recessive
b. sex-linked dominant
c. sex-linked recessive
d. autosomal dominant
e. Y-linkage

Topics for
discussion

1. What is meant by genetic equilibrium?
What factors disturb the genetic equilibrium of a
natural population?
Which factors are more influential in large versus
small populations?

2. A population is in genetic equilibrium for the
polymorphic genes B and b (B dominant). What are
the gametic and zygotic frequencies of B if the
allele b is homozygous in 49 percent of the
population?

3. What are the minimal different cellular components that a bacterial spore must contain? How
does this compare with a plant seed?

4. Several external factors are able to influence
the polarity of rhizoid formation in the embryo of
the seaweed Fucus. For example, a rhizoid (primary root) is formed on the warm side of a temperature gradient, on the acid side of a pH gradient,
and on the shaded side of a light gradient. The
establishment of polarity requires only a short
exposure to the gradient. Suggest an explanation
for this effect.

5. Phenylketonuria is caused by a recessive allele
that prevents phenylalanine from being converted
to tyrosine. It has been reported that it is
possible to "cure" phenylketonuriacs by feeding
them a phenylalanine-free diet.
a. Would this "cure" modify the contention that
the disease is caused by a recessive allele?
b. Would the "cure" affect the frequency of the
recessive allele in future generations?

Topics for
discussion

1. Distinguish between determination and differentiation in development.
What kinds of experiments can be used to show that
the developmental fate of a given cell is already
determined?

2. Why is gastrulation a critical event in animal
development?
What factors might be responsible for the differential cell migration that occurs during gastrulation?
What experiments show that the polarity of gastrulation in amphibians is influenced by morphological
determinants in the fertilized egg?

Written problem

The enzyme lactic dehydrogenase, or LDH, present
in all cells of almost every animal, consists of
four subunits. These can be all identical or not;
when they are not identical, they are of two similar types distinguished only by electrophoretic mobility. The enzyme may consist of any combination
of type A and B subunits, so that there can be in
any cell a mixture of five enzyme forms:
A_4; A_3B_1; A_2B_2; A_1B_3; B_4. All are functional.

This situation holds even in inbred mouse lines.
The ratios of these five forms in a given animal
vary from tissue to tissue and during development.
For example, LDH of mouse eggs is all B; mouse
embryo organs at the 9th day have LDH with many
more A than B subunits; adult mouse heart muscle
has 90 percent A_1B_3 enzyme.

a. Are A and B subunits the products of two separate genes or of two alleles of the same gene in
two homologous chromosomes? Why?

b. What do the differences in LDH composition in
embryogenesis and in different tissues tell you
about gene action in development?

c. Two inbred lines of mice exist whose B subunits
of LDH can be distinguished electrophoretically:
B^x and B^y. Using <u>allophenic</u> mice derived from
these two lines, you find that the LDH in their
heart cells (uninucleate) contain only $A-B^x$ or
$A-B^y$, whereas LDH in muscle fibers (polynucleate)
contain all combinations of $A-B^x-B^y$.
What do you conclude as to the events that generate
muscle fibers?
Do these findings suggest in which compartment of
the cells the LDH protein is assembled?

Topics for
discussion

1. Various embryonic tissues can be dissociated
into single cells by shaking in a low salt solution containing trypsin. Cells were obtained in
this way from embryonic chick liver, cartilage and
epithelium. Under suitable conditions, these
cells will reaggregate in specific ways. Liver
cells tend to aggregate with liver cells, cartilage
with cartilage, etc. When two cell types are
allowed to aggregate in the presence of each other
the following aggregates are formed:
inner mass of cartilage cells, covered with liver
cells;
inner mass of cartilage cells, covered with epi-
thelial cells.
a. What factors might mediate tissue-specific
aggregations?
b. What factors might determine the patterns of
aggregation when two different cell types are
allowed to aggregate in the presence of each
other?
c. When a mixture of liver, cartilage, and epithe-
lial cells reaggregate together, what patterns of
aggregation might be formed?
d. What would the formation of the aggregate cartilage (inner), liver (middle) and epithelium (outer)
tell you about the type of interactions discussed
in part (a)?
e. Why are specific tissue-tissue interactions important in development?

2. The following experiment was done using two
species of Acetabularia which differ in the shape
of the cap. Species X = normal, round cap;
species Y = spiked cap. A capless, enucleated
stalk from X is grafted to a rhizoid (contains
nucleus) from species Y. After a few months, a
cap is regenerated--its form is neither round nor
spiked, but is intermediate. This cap is removed
and a second cap regenerates. This cap is spiked.
a. Why is the first regenerated cap intermediate
in shape?
b. Why is the second cap spiked?
c. How could you test your hypotheses?

3. When normal ectoderm from a nonpigmented chick
embryo is transplanted over mesoderm of a wing
bud from a polydactylous, pigmented chick embryo,
a nonpigmented wing is formed which has more than
the normal number of digits. When the reciprocal

experiment is done (polydactylous, pigmented ectoderm + normal, nonpigmented mesoderm) the resultant
wing is pigmented and has the normal number of
digits.
a. Propose an explanation to account for these
findings.
b. How could you show that either tissue-tissue
contact or the passage of either small or large
molecules is required for the induction of wing
formation?

4. Two mutants A and B of the slime mold Dictyostelium were isolated. Both lack the ability to
aggregate. Antibodies made to normal Dictyostelium
react with antigens of the cell surface of normal
slime molds and to those of mutant A, but not to
those of mutant B. Normal slime mold cells migrate
towards a source of cyclic AMP, but mutants A and
B do not.
What mutations might be responsible for the failure
of strains A and B to aggregate?

Topics for
discussion

1. What is a hormone? What are the essential
elements in a hormone-mediated regulatory system?

2. What series of events occur in response to a
decrease in the level of blood glucose?

3. The female contraceptive pill contains progesterone and estrogen in the amounts corresponding
to the levels just before menstruation. How does
this prevent pregnancy?

4. Suggest a formula for a male contraceptive
pill, based on what you know about male and female
hormones.

5. What are the roles of ATP and creatine-phos-
phate in muscle contraction? How do muscle cells
convert chemical energy into mechanical energy?
What is the role of Ca^{++} ions in muscle contraction?

Topics for
discussion

1. Compare the information content of the immuno-
logical system and that of the nervous system.
Does this question make any sense?

2. What characteristics distinguish a chemical
synapse from an electrical synapse? For what
circuits would you prefer one over the other?

3. Why is it important that each type of cell produces only inhibitory or only excitatory synapses?

4. Why are γ efferent fibers important in maintaining the signal from a stretch receptor as the
muscle shortens?

5. What is the difference between local potentials (LP) and action potentials (AP)? Do LP
play a role in propagating AP?

6. What is a saltatory potential and why does it
arise in myelinated fibers?

7. Convince yourselves of the truth of this state-
ment: the membrane potential is the key to the
whole set of events in nerve conduction, because
it determines the permeability of nerve membranes
to ions.

8. A miniature end plate potential, like a mini-
skirt, releases one tease at a time; an action
potential to a chemical synapse "goes all the way."
Why?

9. A limb bud implanted on the back of a salaman-
der embryo develops into a limb innervated by the
local spinal cord and ganglia. Yet, when the extra
limb is stimulated, one of the normal limbs
twitches. What may this mean?

10. Why does a mirror invert objects left-to-right
but not up-to-down?

11. What is the common process in all sensory
transduction (stretch, light absorption, sound
reception, etc.)?

This section provides some examples of the type of questions used in examinations (all open-book type).

Question 1

Hemoglobin, the major protein of red blood cells, has one important function besides the ferrying of oxygen. It serves as a buffer, contributing to keep the pH of blood near pH 7.0.

1. What properties of the hemoglobin molecule contribute to its buffering action?

2. Which amino acid would be most responsible for the buffering action, and why?

3. When hemoglobin is attacked by the enzyme trypsin, the overall electric charge at pH 7.0 remains approximately the same. The total number of charges, however, as determined by the amount of acid or base needed to neutralize all charges ("titration") is increased many times. Explain.

4. If you remove trypsin (or add an inhibitor of trypsin) the effects of the enzyme are not reversed. Why?

Question 2

You are given a solution of an "unknown" substance which, on test, proves to have a single pK_a at 10.79. Upon treatment with the enzyme trypsin, which breaks polypeptides on the -CO... side of either lysine or arginine, the substance reveals two new acid dissociations with $pK_a = 2.17$ and 9.75, respectively.
Upon complete hydrolysis each mole of the original substance yields one mole of lysine, one mole of leucine, and four moles of threonine.
What is the structure of your unknown substance? (Explain if necessary).
How would you separate the amino acids in question (lysine, leucine, threonine) by electrophoresis? (Indicate conditions used, and why. A hint: choose pH corresponding to one of the pK_a.)

Question 3

Glycogen is a polymer made up of glucose units: glycogen = $(glucose)_n$. The synthesis of glycogen involves the following reactions:

glucose-6-phosphate $\rightarrow$ glucose-1-phosphate

$\Delta G_0'$ +1.7 kcal/mole

glucose-1-phosphate + ATP $\rightarrow$ ADP-glucose + pyrophosphate

$\Delta G_0'$ +0.5 kcal/mole

ADP-glucose + (glucose)$_n$ → (glucose)$_{n+1}$ + ADP

$\Delta G_0'$ -3.2 kcal/mole

1. What is the ΔG_0 of the overall series of reactions as written? What would be the equilibrium constant (approximately)?

2. In cells or cell extracts the reaction proceeds to completion in the direction of glycogen synthesis. Is there any reaction that should be added and that would cause the equilibrium to shift in favor of extensive glycogen synthesis? Why?

3. In inserting one molecule of <u>free</u> glucose into glycogen, how many "high energy" phosphate bonds are used up? Explain.

Question 4

1. A culture of yeast is growing anaerobically with glucose-5-^{14}C (that is, glucose labeled in the C atom in position 5) as the only source of carbon. Where will the label be found in the following compounds?
a. ethyl alcohol
b. glycerol ($CH_2OH-CHOH-CH_2OH$; mainly present in membrane phospholipids)
c. long chain fatty acids ($CH_3-(CH_2)_n-COOH$; made by repeated additions of C-C groups derived from acetyl-SCoA)

2. When the anaerobically growing yeast culture makes fatty acids from acetyl-SCoA, it produces proportionally less CO_2 and ethanol from glucose than if it fermented glucose without growing. This creates an apparent difficulty for the functioning of the glycolytic pathway. Can you see what the difficulty is, and figure out how the growing yeast cells solve it? (Hint: one ATP and 2 NADPH or NADH molecules are required to add a C-C unit from acetyl-SCoA to a growing fatty acid chain.)

3. To grow anaerobically on glucose, yeast must also make glycerol (see above) from dihydroxyacetone-phosphate. Does this present the same problem as the making of fatty acids?
Why, or why not?

Question 5

A strain of <u>E. coli</u> has a mutation that blocks the "condensing enzyme," which catalyzes the entry reaction into the Krebs cycle:

oxaloacetic acid + acetyl-SCoA → citric acid.

This mutant cannot grow on glucose as the only C source.

1. What else will it need, and why?

2. Those glucose carbon atoms that are not incorporated into cell substance are excreted by this mutant as CO_2 and acetic acid. Why acetic acid and not acetyl-SCoA? Does the cell get any energy by splitting acetyl-SCoA?

3. Where would the carbon atoms from glucose go if the condensing enzyme were not blocked (that is, in a normal cell)?

4. If normal cells growing in nonlabeled glucose receive some $^{14}CO_2$, part of the ^{14}C ends up in proteins, etc. If amino acids are present, this incorporation of ^{14}C is suppressed. Explain. ($^{14}CO_2$ is added as bicarbonate HCO_3^-, which releases CO_2: $HCO_3^- \rightleftharpoons OH^- + CO_2$.)

Question 6

The incorporation of an amino acid (for example lysine) into proteins is mediated by the following reaction:

(1) lysine + ATP + tRNA → lysyl-tRNA + AMP + 2 iP.

This reaction has a $\Delta G_0'$ of -7.5 kcal/mole.

If you measure the K_{eq} for the reaction

(2) lysine + ATP → lysyl-AMP + iPP,

which is catalyzed by the "activating enzyme" you find a value $K_{eq} = 0.2$.

1. What is the $\Delta G_0'$ for reaction (2)?

2. How do you explain the difference between the $\Delta G_0'$ values for reactions (1) and (2)?

3. If you double the concentration of activating enzyme, what happens to K_{eq}? to the rate of the reaction?

4. If you add a lysine analog that inhibits competitively the activating enzyme, what happens to K_{eq}? to the reaction rate?

Question 7

1. Certain substrates labeled with ^{14}C are incubated with a muscle extract in the presence of air. Malonic acid, an inhibitor of succinic dehydrogenase (the enzyme that converts succinic to fumaric acid), is added in excess. After a certain amount of substrate has been used, various Krebs cycle intermediates are isolated and analyzed.

Substrate	substance analyzed
*CH_3COOH (acetic acid)	citric acid
$CHO-CHOH-*CH_3OPO_3^{--}$ (glyceraldehyde-3-phosphate)	α-ketoglutaric acid
$CH_2OH-CO-CHOH-CHOH-*CHOH-CH_2OPO_3^{=}$ (fructose-6-phosphate)	succinic acid

Predict the positions of the ^{14}C atoms in each of the various substances (draw formulas and explain when necessary, * indicates ^{14}C).
What would happen if the inhibitor, malonic acid, were not added? Where would the ^{14}C ultimately end up?

2. Malonic acid is a competitive inhibitor of succinic dehydrogenase. Recall that for competitive inhibition

$$V = \frac{V_{max}}{1 + \frac{K_m}{[S]}(1 + \frac{[I]}{K_I})},$$

Consider three cases: $[I] = 0$; $[I] = K_I$; $[I] = 9K_I$.

What will be the concentration $[S]$ of succinic acid (in multiples of Km) to give 1/2 maximal reaction velocity in each of the three cases in the presence of the pure enzyme?
Draw curves representing approximately the velocity V of production of fumaric acid as a function of succinic acid concentration for the three above cases.

Question 8

There exist bacteria (called sulfur bacteria) that can grow using only CO_2, $-NH_4^+$, and H_2S as sources of carbon, nitrogen, and energy. Which pathways must these organisms use--
1. to make organic carbon compounds from CO_2?
2. to obtain ATP?
What else would they need besides ATP to make organic compounds from CO_2, and why?
One group of these bacteria can grow without oxygen provided it is given nitrate ($-NO_3^-$). How do they work it?
[Hint: H_2S can be oxidized to SO_4^{--} (sulfate). Compare with photosynthesis.]

Question 9

In E. coli there are two enzymes which in the test tube can make some RNA. One enzyme, RNA polymerase, carries out the following reaction:

(1) nATP + nGTP + nUTP + nCTP $\rightleftharpoons$ "RNA" + 4n iPP
(= pyrophosphate) ($\Delta G_0' = +0.5$ kcal/mole).

The other enzyme, called phosphorylase, catalyzes the reaction

(2) nADP + nGDP + nUDP + nCDP $\rightleftharpoons$ "RNA" + 4n iP
($\Delta G_0' = +1.5$ kcal/mole)

In vivo, RNA is made by reaction (1). What is energetically superior (in vivo) in reaction (1) over reaction (2)?

Reaction (2) is supposedly used by E. coli to break up used RNA and make the pieces available again. Does this seem reasonable, and why?

Would breakdown of RNA by reaction (2) be preferable for the cell to breakdown by the reaction

(3) "RNA" + H_2O $\rightarrow$ nucleotides

and why?

Question 10

Three mutant genes are used in a Drosophila cross:
e eagle (wings spread)
h hairy (extra hairs on head)
f frizzled (hairs turned inward)
The cross +++/ehf x ehf/ehf (+ meaning normal alleles) yields the following progeny:
40 +++
42 ehf
 5 e+h
 7 +h+
 2 e++
 2 +hf
 1 ++f
 1 eh+

1. Which groups represent double crossovers?

2. What is the sequence of the three genes?

3. Give the distance in map units between

(1) h - f; (2) f - e; (3) h - e.

4. Compare the measured distance to the distance obtained from the mapping function. Why is the measured distance less than the calculated distance?

5. Sketch your interpretation of the meiotic events leading to this recombination.

Question 11   In the mouse, animals carrying a dominant gene
C are pigmented, whereas cc animals are albino
(no pigment). Another gene B determines the difference between black (B) or brown (bb). Thus
BBcc is albino, bbCC is brown, BbCc black, etc.

1. If the C and B genes are carried on different
chromosomes (unlinked), what are the genotypes
and phenotypes in the F_2 generation from a cross
CCBB x ccbb?

2. If C and B were linked, what would be the major
genotypes and corresponding phenotypes in the F_2?

3. If you obtain from the cross in (2) some F_2
animals with brown phenotypes, what are their genotypes and how did they arise?

4. Suggest an explanation for the physiological
interaction between genes B and C.

Question 12   E. coli has a mating system that leads to
genetic recombination and permits mapping. Suppose you are given a research project to look for
a mating system in the bacterium Hypotheticus nonexistent.

1. Outline how you would proceed to set up your
tests (strains you would prepare, search for
genetic recombinants, test for polarity as in F^+
versus F^-, etc.).

2. If you got such a system, how would you try to
test if there is a minimum frequency of recombination? What relation would this have to the genetic code?

3. Suppose you could cause E. coli to mate with
H. nonexistent. Could you use this system to test
whether the genetic code is the same? If so, by
what tests?

4. If the genetic code is the same, what does this
mean in terms of the various components of the protein synthetic system?

Question 13   When bacteriophage T4 infects cells of E. coli
the synthesis of all coli proteins stops within 5
minutes. Then as phage proteins begin to be made,
one of them, the main "head" protein, is in such
excess over the others that one can study it almost as if the others were not there. This can be
used to study the effects of various mutations on
a single protein.

A series of mutants with "conditional mutations" in the gene corresponding to the T4 head protein are isolated; some grow only below 28°C; others only in special host bacteria. When the head protein is extracted from bacteria infected under conditions where phage does not grow (nonpermissive condition), one finds that in some mutants the head protein is made, is of full length, but is changed in amino acid composition; in others, the protein chain is variously shortened.

1. Interpret these results, explaining what would happen in the permissive condition, where the phage mutants do grow.

2. The above mutants are used pairwise to test (a) for complementation, and (b) for recombination. How are the tests done? What results would you expect in each test, and why?

3. In one specific case you study mutant X and mutant Y, whose defective proteins differ from the normal one as follows:
Normal: ... tyr-glu-leu COOH
X   ... tyr-arg-leu COOH
Y   ... tyr COOH
Can you find recombinants? Can you figure out what the mutations were at the level of the mRNA (genetic code)?

Question 14

E. coli can use the sugar arabinose as a food because it has three enzymes, A, B, and C, which acting in stepwise fashion convert arabinose to a splittable sugar.
The genes corresponding to the three enzymes A, B, C are linked sequentially in the E. coli genetic map.
Addition of arabinose to a culture growing in meat broth increases many times the rate of synthesis of enzymes A, B, C compared with the rate of overall protein synthesis.
A mutation S → s, not in gene A, B, or C, causes the synthesis of these enzymes not to respond to arabinose. [Note that this system differs from the lactose system!]

1. By what experiment on E. coli cells can you verify the linkage order of the genes A, B, and C? (Be brief but specific; use sketch if you wish.)

2. What happens to the rate of enzyme A synthesis if you remove very quickly the arabinose from the medium? Why?

3. Do you expect arabinose to increase the rate of synthesis of each of the three enzymes A, B, and C by the same numerical factor? Why, or why not?

4. An experiment was done to explain the nature of mutations. The following partial diploids were prepared:

(1) s Abc/S aBC

(2) s Abc/s aBC

Diploid (1) responds to arabinose by making A, B, and C in maximal amounts. Diploid (2) does not respond to arabinose.
Explain the results in terms of the theory of regulatory proteins (repressor and activators).

Question 15   The following pedigree is for genetic deafness in a group of families (deaf individuals = shaded). The two input families a and b were not related.

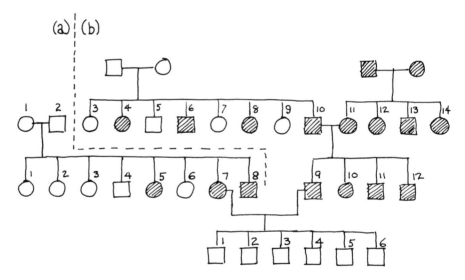

1. What kind of genetic defects are involved? (dominant or recessive? in the same or different genes? sex-linked or autosomal?)

2. What are the likely genotypes of individuals 1 and 2 in generation II?

3. What proportion of the individuals in generation IV could you expect to be heterozygous for one or more deafness gene? Why?

4. What analogy is there between a situation of this kind, as you explained it above, and complementation as tested with partial diploids of E. coli or with mixed infection with phage mutants?

5. Deaf people are less than 0.1 percent in the population. Is there anything in this pedigree to suggest that the genes for deafness would not obey the Hardy-Weinberg equilibrium in the human population? What kind of deviation would it be?

Question 16

In man the genes for color-blindness and hemophilia are both on the X chromosome with about 10 percent crossing-over. A pedigree involving both traits is given below:

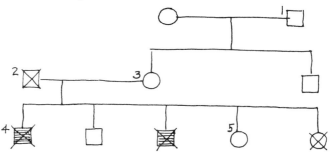

If a symbol is crossed (X), the individual is color blind. If the symbol is shaded (≡), the individual is hemophiliac. Otherwise the individual appears normal.

1. Is the allele that produces hemophilia dominant or recessive?

2. Is the allele that produces color blindness dominant or recessive?

3. The individuals numbered 1, 2, 3, and 4 are among those in the pedigree whose genotype can be determined uniquely. Write the genotypes of these 4 individuals, using the following symbols:
H, h the 2 alleles for the hemophilia locus,
B, b the 2 alleles for the color-blindness locus,
———— an X chromosome,
———→ a Y chromosome.
For example, a male who received a Y chromosome from his father and an X chromosome bearing H and b from his mother would be written

H⎯⎯b

⎯⎯⎯→

and a female who received the 2 dominant alleles H and B from one parent and the 2 recessive alleles h and b from the other would be written

H⎯⎯B

h⎯⎯b

4. What is the probability that a son of the woman marked 5 will be color-blind?
What is the probability that her son will be hemophiliac?

Question 17   There exists a mutant n of the toad <u>Xenopus</u> whose cell nuclei lack a nucleolus (the organelle where rRNA is made). The mutation is recessive lethal.

1. How can a recessive lethal mutation be maintained in existence? What is an analogous situation in man or some other mammal?

2. How can the heterozygotes Nn be distinguished from the homozygotes NN by simple microscopic observation? by nucleic acid hybridization? by genetic crosses?

3. Homozygous nn animals live to the free-swimming tadpole stage, but then die. What does this tell you about maternal inheritance?

4. Compare the bearing of this case, of human galactosemia, and of Huntington's chorea on the question of genetic control of development. (Feel free to write a brief essay.)

Question 18   According to current theories of muscle contraction, the interaction between components of the contractile mechanism can occur only when the concentration of Ca^{++} ions has a precisely specified value.

1. Sketch a rough diagram of the organization of the striated muscle fibers, indicating--
a. the contractile components and their relative motions;
b. the location where Ca^{++} may be reversibly stored or released;
c. the way an action potential coming from a neuromuscular junction may affect Ca^{++} movement specifically.

2. During contraction ATP is converted to ADP + H_3PO_4. ATP can be regenerated by three mechanisms. Mechanism 1: ADP + phosphocreatine $\rightleftharpoons$ ATP + creatine. The equilibrium of this reaction is toward the right. What type of compound (in terms of free energy of hydrolysis) will phosphocreatine be, and why?

3. What are the other two mechanisms that generate ATP in the muscle, and what are their relative contributions in different situations?

Question 19

During pregnancy, embryonal cells (trophoblast) in the placenta secrete <u>chorionic gonadotropin</u> (CG), which stimulates the corpus luteum in the ovary to secrete the very high concentrations of estrogen and progesterone needed to maintain pregnancy. CG production is not controlled by hypothalamus and pituitary.

1. How do estrogen and progesterone act in maintaining pregnancy?

2. How is ovulation prevented during pregnancy?

3. Could both goals--high estrogen and progesterone, and no ovulation--be compatible if the normal regulatory system were used (that is, instead of CG)?

4. Consider the following embryonal products or constituents:

a. chorionic gonadotropin;
b. histocompatibility antigens on the embryonal cells of the placenta;
c. Rh antigen on red blood cells of the embryo.
Each of these interacts in some way with the mother's organism. State briefly how and why.

Question 20

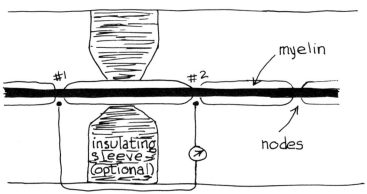

The above experimental set-up is used to measure and, when desired, to block external currents that may pass along a myelinated nerve fiber.

1. Why, in the <u>absence</u> of the insulating sleeve, does the external system register passage of current when a "saltatory" action potential goes through the nerve?

2. If the external conductivity is reduced to near zero by the presence of the insulating sleeve, the action potential fails to go through. Why?

3. In the absence of the insulating sleeve, as an action potential reaches from node 1 to node 2, the current measured at node 2 first flows outward, then turns inward. How do you explain this in terms of the events (depolarization, etc.) at the node?

4. Cocaine is a drug that allows passive ion flows in nerves, but blocks changes in membrane potential and, therefore, excitability of nerve fibers. If cocaine is applied to node 2, only the outward current is observed, but not the inward one. Does this fit your answer to (c) above? If not, think again.

Question 21

1. You have a muscle-spinal cord system as that shown on page 326. If you were to bring the interneuron •————⊣ on the left of the sketch to a membrane potential E_m = -120 mV, how would this affect the response to stimulation of the flexor muscle? Why?

2. The transmitter acetylcholine is made in the nerve ending of a neuromuscular junction (reaction 1) and is broken down outside (reaction 2) in the synaptic gap.

(1) choline + acetyl-SCoA ⇌ acetylcholine + HSCoA

(2) acetylcholine + H_2O ⇌ choline + acetic acid

Choline is not made in the cell: it comes from outside. Its internal concentration is 1,000 times higher inside than outside.

What is the minimum number of ATP equivalents that would be used by a nerve cell to produce one mole of acetylcholine? Why? (Think!)

3. The motor axon α in the figure on page 326 makes acetylcholine synapses to many muscle fibers. The poison curare injected in the blood abolishes muscle contraction in response to stretch receptor stimulation. Assuming you can do any measurement you want on the elements of the circuit, how would you test whether curare acts at any one of the seven or more possible sites that may be blocked by curare? (List them and say how you might test if they are blocked).